Climate Change, Energy, Ecology, Health

Also by Emil Morhardt

Climate Change, Energy, Ecology, Health

J. Emil Morhardt, Editor

CloudRipper Press

Cutting Edge Books

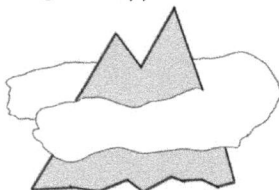

CloudRipper Press
Santa Barbara, California
www.CloudRipperPress.com

Morhardt, J. Emil
    Climate Change, Energy, Ecology,
    Health/ J. Emil Morhardt, Editor.

ISBN 978-0-9963536-6-3 (paper)

Table of Contents

Forward

J. Emil Morhardt

While this book was being written, President Trump, like most of the other Republican candidates for President denied loudly that climate change is happening, and the few who acknowledged it were quite sure that humans had no part in it. As of this is being published, Scott Pruitt, the EPA administrator under Trump still denies any human contribution. It is too bad that they are so scientifically illiterate. Low-information candidates and voters make bad decisions, and the World is in a situation where good decisions are very much needed. Reading this book will significantly increase your personal scientific literacy on the topic, and perhaps convince you not to vote for climate change deniers.

The focus of this book is the interactions between climate change and energy production, ecology, and human health, as well as a few of the responses of humanity to these interactions. It is not a textbook, but a series of chapters discussing subtopics in which the authors were interested and wished to write about. The basic material is cutting-edge science; technical journal articles published within the last year, selected for their relevance and interest. Each author selected eight or so technical papers representing his or her view of the most interesting current research in the field, and wrote summaries of them in a journalistic style that is free of scientific jargon and understandable by lay readers. This is the sort of science writing that you might encounter in the New York Times, but concentrated in a way intended to give as broad an overview of the chapter topics as possible. None of this research will appear in textbooks for a few years, so there are not many ways that readers without access to a university library can get access to this information.

One place is scientific blogs on the Internet, and most of the material in this book will appear in the blogs ClimateVulture.com and EnergyVulture.com by mid-2018, but all of the material is available here.

This book is intended be browsed—choose a chapter topic you like and read the individual sections in any order; each is intended to be largely stand-alone. Reading all of them will give you considerable insight into what climate scientists concerned with energy, ecology, and human effects are up to, and the challenges they face in understanding one of the most disruptive—if not very rapid—event in human history; anthropogenic climate change.

Section I—Managing Natural Resources and Agriculture

Community-Based Natural Resource Management Methodologies

Deedee Chao

Community-based natural resource management (CBNRM) is an approach to natural resource management (NRM) that combines the goal of conserving natural resources with the goal of social and economic benefits for the local communities in which these natural resources are located. By addressing the environmental, social, and economic issues within social-ecological systems (SES), CBNRM aims to intertwine self-interest with environmental interests so that local communities can continue to carry out sustainable NRM practices without external intervention. Many communities lacking in resources and infrastructure tend towards community-based strategies to address issues by themselves, as these are more accessible and natural methods for smaller, more traditional communities. These strategies can be expedited and improved with aid from outside sources, such as non-governmental organizations or governmental agencies. However, difficulties in communication and collaboration between these different stakeholders who all have varying goals and levels of commitment make it complicated for CBNRM to be effectively carried out. In order to have successful outcomes from CBNRM, it is imperative that local communities are able to communicate their needs and realities to external parties, and that external parties can pinpoint local needs and address them.

Under the effects of climate change, in a time of increased concerns around sustainability, CBNRM has become increasingly relevant as an adaptive approach, because many of the areas and communities most adversely affected by extreme weather patterns and related problems are remote, rural communities that often lack monetary resources, educational backgrounds, and proper infrastructure– exactly the kind of communities for which CBNRM is most suitable. While approaches via policy and scientific advancements are imperative to addressing climate change issues, these developments are not always available or accessible for disadvantaged communities. As a result of CBNRM's pertinence to the global problem of climate change, environmental literature around different CBNRM methodologies and practices and their effects on local communities has been growing, as has literature that attempts to fit theoretical frameworks to CBNRM's complex operational structures in order to increase understanding and facilitate effective implementation. This chapter and its compilation of recent scientific articles on CBNRM and related issues aims to document various approaches to CBNRM around the world, their effectiveness, and supporting theories and frameworks for continued usage.

The chapter begins with a case study on the documented effects of extreme weather events and erratic weather patterns on remote, agricultural communities in the Himalayas, demonstrating that climate change has indeed had and continues to have a pronounced impact on various places around the world. It then moves on to a couple of examples of these remote communities moving towards community-based adaptive strategies of their own accord, in order to survive the effects of climate change: farmers in the Himalayas implementing community-based agricultural practices, and indigenous communities relying on community-based adaptive strategies to survive extreme weather events in central Australia. With a foundation in the effects and natural responses to climate change established, we move outside of these local communities to looking at how changes at the policy level can affect and often exacerbate local problems, when the government is out of touch with the needs of the people, using Australia again as a case study. The importance of communication and cooperation between local peoples and governmental entities is further highlighted through a paper on why formalized social networks between citizens and the government were imperative for co-ordinated urban forest management in northern Japan.

From there, an article on community-based environmental management (CBEM) through the lens of multi-level governance (MLG) theory grounds the aforementioned issues and components of CBNRM in theoretical frameworks, providing context for how to better think about the multiple stakeholders and effective methods of collaboration. A nationwide study on community-based rangeland management (CBRM) in Mongolia then documents an example of successful CBNRM achieved by local herders working with international NGOs, showing that CBNRM can be effective for social outcomes, and potentially effective for economic outcomes, when implemented correctly. Finally, a comparative analysis of Latin American communities using prospective structural analysis (PSA) as a tool for implementing and analyzing CBNRM provides some insight into how to best proceed with CBNRM theory and practical application.

Community Perceptions and Responses to Climate Change in the Himalayas

In the Hindu Kush Himalayas (HKH), natural disasters related to climate change have increased over the past few years, leading to long dry spells, flash floods, landslides, and more across the region. Coupled with increasingly erratic rainfall, these occurrences have made it necessary for local mountain communities to increase their adaptive capacity in order to survive. In order to have effective adaptation strategies, it is essential to have an understanding of the impacts of and responses to climate variability, which can only be attained through a thorough information base. However, in rural areas of developing countries such as the HKH, it is difficult to have reliable information, which creates a challenge for groups such as governmental agencies and local civic organizations that seek to address challenges of climate change. To this end, the International Centre for Integrated Mountain Development (ICIMOD) carried out a participatory assessment in 2010–11, documenting community perceptions, community resources, and institutional dependency in relation to climate variability. This assessment spanned 90 villages in the HKH, across 15 districts of Bhutan, India, and Nepal, within an altitudinal range of 50–3500 MSL. Pandit *et al.* (2016) highlights the impacts and responses of these mountain communities in order to analyze and draw conclusions from the data, as well as to determine methods of bettering adaptive strategies.

The participatory assessment was conducted via the Participatory Rural Appraisal (PRA) toolkit, comprising five tools for different measurements: Seasonal Calendar, to document weather trends; Hazard Ranking, to rank weather events on how hazardous they were to local livelihoods; Seasonal Dependency Matrix, a Seasonal Calendar to map how communities depend on different support systems; Seasonal Activity Calendar, to document farming and non-farming activities throughout the year; and Institutional Dependency, a Venn Diagram to document community dependencies on various institutions. In order to better identify impacts of climate change and community responses, the PRA toolkit process was conducted again based on the original Hazard Ranking results, with special emphasis on the findings of the Seasonal Calendar and Seasonal Activity Calendar.

All communities in the study area perceived weather pattern changes, especially changes in precipitation. Almost all areas experienced a reduction in the annual duration of rainfall, with this reduction being most apparent in western locations, producing a statistically significant trend longitudinally. There was also a reduction in the duration of annual snowfall, with a complete absence of snowfall in some places. Coupled with the reduction in rainfall was an increase in the duration of dry periods across the board, particularly in Uttarakhand and Nepal. Trend analysis of the duration of dry periods also showed western locations having significantly longer dry periods. Although the majority of the HKH region lacks sufficient data on annual precipitation over the past century, other data show that the frequency of heavy rainfall events is increasing while lighter rain events are decreasing in South Asia generally, which correlates with community perceptions of weather events.

As a result of weather variability, agriculture, and thus income and livelihood, have been adversely affected. Because of varied precipitation patterns, crops have performed poorly or failed altogether, making agricultural households extremely vulnerable. Further, precipitation patterns have also led to depleted soil moisture and overall water scarcity, creating a long-term problem that communities have coped with by increasing the work burden on women, who are in charge of collecting water. With a longer, more difficult workday, women have been exposed to more health issues. In addition, men have been forced to move outside of their immediate communities to look for seasonal wage employment with nearby coal mines, further increasing the burden on women to carry out work at home and take on traditional men's jobs such as jungle clearing. Overall, the changes in weather have led to various snowballing effects that endanger the livelihoods of local communities and make them even more vulnerable to chronic poverty.

The researchers split up responses to climate variability into two categories: coping responses, which are reactive, temporal, and immediate; and adaptive responses, which are anticipatory, precautionary, and long-term. Many farmers replaced their crops in order to adapt to shifts in climate. In Uttarakhand and Nepal, the majority of farmers chose to replace subsistence crops with cash crops. Although they named weather changes as a reason for this shift, there is also a growing market opportunity for cash crops that could be another reason for the shift. Meanwhile, in Nepal, Bhutan, and Northeast India, farmers have been more focused on using traditional varieties of food crops as replacements, rather than moving to commercial crops. A possible reason could be that there is easier access to new crops in Uttarakhand, while these are less accessible in areas like Bhutan and Northeast India. Communities have also been diversifying their agricultural practices in order to become more resilient, often phasing into raising smaller livestock rather than large livestock. When faced with decreased rainfall and less time to dedicate to tending animals, small livestock that require less food and maintenance, such as chickens and goats, are more preferable. To address water scarcity in the long term, several communities have constructed water harvesting structures, while others have moved towards oak forest regeneration and other natural catchment conservation efforts. Still others have dedicated efforts to reviving traditional springs and water sharing practices. Outside of farming, households have moved towards non-farm opportunities, such as mining and other daily or seasonal wage opportunities. In more extreme cases, where crop failures or large loss of livelihood has occurred, families have moved into urban areas to seek other employment opportunities. This has been the case especially in Nepal. All in all, coping strategies that have been successful have been transitioned into correlating adaptive strategies, all meant to address the issue of survival via traditional or newer livelihood practices.

The PRA assessment also looked at institutional dependencies and found that communities first and foremost relied on informal institutions like friends and family, which are the most accessible and critical for instant relief. From there, communities would move on to relying on formal civic institutions, such as the church. While formal institutions such as government agencies had the most extensive services, these were often difficult to access and had barriers to communication and information flow, rendering them unhelpful for communities in immediate need.

As a whole, adaptive responses to the effects of climate change have been aimed at reducing community vulnerability and increasing communities' abilities to handle stressors. This can be done by increasing the resilience of livelihood systems and thus food and income security, either by addressing production systems, or changing crops, resources, or occupations. The researchers also remind us that it is important to note that climate change is not the only cause of vulnerability and risk in mountain communities: resource degradation, changes in land use, and losing touch with traditional practices and ways of life are also issues that must be considered if effective adaptive methodologies are to be developed and implemented. In order for support systems to be more helpful to local communities, government agencies at the local level should work to become more accessible and responsive. Communities can be aided by interventions in natural resource management methodologies, introductions to more resilient food crops, and stronger support systems, especially in terms of better governmental response rates and credit lending programs.

Facing Climate Change: Community-Based Strategies for Himalayan Hill Agriculture

Climate change has been having adverse consequences worldwide, but in more ecologically sensitive mountain regions such as the Himalayas, these consequences become even more pronounced, thanks to rapid altitude changes that result in varying temperatures, rainfall, flora, and other natural phenomena. In Uttarakhand, the region explored in Jethi *et al.* (2016), the Himalayas range from 300 to 3,600 meters, with habitats from snow-covered peaks to fertile valleys and dense forest cover. This topography makes the region particularly susceptible to climate change. Further, roughly 70% of the population lives in rural areas and primarily depends on agriculture as its livelihood, although only 14% of the region is under cultivation. For the past several years, Uttarakhand has experienced a number of natural disasters such as flash floods that caused a high loss of human lives, regional infrastructure, and personal property. This is largely because of the effect of climate change on glaciers, which melt more quickly, leading to a number of water-related issues, particularly temperature shifts, changes in rainfall patterns, and changes in the amount of snow cover.

The higher temperatures from climate change not only impact glaciers, but the resultant freshwater temperatures, which have risen and allowed for more microbes to breed in the water, in turn adversely affecting those who drink the water. In addition, precipitation is coming increasingly in the form of rainfall rather than snowfall, and the erratic and intense pattern of rainfall has caused landslides, flash floods, and other events that damage agricultural practices. Since the rainfall is largely concentrated in a few months of the year, and snowfall has decreased, the region also suffers drought as no water is released throughout dry months. A lack of snowfall and gentle rainfall also prevents proper irrigation of the soil, which is important for successful agriculture. As a result of these changes, locals have had to largely give up on the cultivation of important crops such as rice, wheat, potatoes, and more.

Climate change has also created problems with wild animals disrupting and destroying crops, as their forest habitat has changed and led to a shortage of food. Insect populations have also changed, leading to an increase in pests that cause crop yield loss but a decrease in pollinators that aid seed formation in cash crops, such as rye, amaranth, sesame, etc. These effects on agriculture place it and climate change into a vicious cycle, where agricultural practices have to be changed because of climate change, leading to increased cultivation areas that lessen forest coverage, which further exacerbates climate change, which leads to the need for more agricultural changes. It thus becomes necessary for agricultural practices to be changed in such a way that the updated methods and practices do not contribute to climate change.

It is also important to consider the consequences for communities that begin to lose their traditional diets and ways of life; in Uttarakhand, this has manifested as a loss of healthy traditional grains and vegetables, and a newfound reliance on markets for food. Further, women who have played a major, essential role in the agriculture of the region have borne the brunt of these changes, as they struggled to deal with the aforementioned problems caused by climate change. Women from the region have traditionally been in charge of managing natural resources, taking care of livestock, and collecting firewood and livestock feed. Because these tasks have become more difficult and cultivation areas have spread out further, the physical exertion involved with both manual labor in the fields and housework at home have led to a number of physical problems, such as lower back pain and infections. These issues are further exacerbated by the lack of access and control women hold over these resources, as well as the marginalization that prevents them from having a voice and a say in how to proceed in the wake of climate change.

As such, Jethi *et al.* (2016) emphasize that it is necessary to recognize the struggles specific to women in climate change case studies, and in this case, center their knowledge and expertise as the community works to survive through the changing climate. The researchers document current measures being taken, such as replacing grain crops with vegetables that have higher production with fewer resources, using a mixed cropping system of multiple crops that prevent mono-crop failure, and terracing fields with trees and plants that can be used as livestock fodder and as firewood. The study also identified a number of medicinal plants that would both sell for high prices and thrive in the current climatic conditions, creating another source of income for the Uttarakhand community. Another recommendation is the creation of irrigation infrastructure for the fields to get enough water, and community water tanks that would make it easier for people to obtain drinking water. The final point stresses that climate change studies and approaches should be multidimensional, taking into account factors such as globalization, poverty, and social stratifications. Overall, this study is valuable because it not only observes climate change effects, but also ways in which the community has responded, and what else it could do from now on, especially taking into account local gender issues.

Investigating Indigenous Community-Based Adaptation in Central Australia

The advance of climate change necessitates not only mitigation strategies, but adaptation strategies to address the "locked in" irreversible effects of climate change thus far. Adaptation becomes especially important in marginalized communities, which are particularly susceptible because of locale and a lack of resources. In Australia, where the average temperature is projected to rise by 1–5°C by 2070, Aboriginal communities located in the remote inland region are notably vulnerable because of poor infrastructure and limited access to essential services. Additionally, indigenous traditional views on health and adaptation in these communities further complicate the research process into best climate adaptation strategies, as they often do not fit into the conventional framework of Westernized health and adaptation concepts. Race *et al.* (2016) seek to investigate the adaptive capacity of two remote Aboriginal communities in central Australia in order to better understand pre-existing adaptive capacities within local communities and how these could blend with innovative, scientific adaptive strategies to best address climate change.

The researchers chose two contrasting locations within central Australia to better study the adaptive capacity of a community in different settings. The first community, Alice Springs, is in a remote location with a population of 25,000 people, 19% of which is Aboriginal and 75% of which speak only English at home. The second community, Lajamanu, is much smaller, with a population of 700 people, 90% of which is Aboriginal and only 9% of which speak only English at home. In Alice Springs, the median weekly household income is $1,676, while Lajamanu's is $1,119, and when assessed for Aboriginal households only, Alice Springs' drops to $1,073 and Lajamanu's drops to $743, demonstrating the socioeconomic disparity between Aboriginal communities and other local inhabitants. Because of climate change, both locations have considerable yearly variation in weather patterns, but Lajamanu's is more pronounced than Alice Springs'. Even more concerning is the estimation that mean daytime temperatures in both locations could rise by as much as 7°C by 2100.

The study goes on to define "adaptive capacity" as the "ability of individuals and communities to cope with change" via resources that will help maintain or enhance livability in new conditions. Such resources are limited not only to physical assets and materials, but can also extend to social networks within the local community, as well as individual attributes such as local knowledge or skillsets. In order to maintain and increase this resource pool, reliable employment and payment is usually necessary so communities can generate enough income for long-term investment in infrastructure, healthy lifestyles, education, and more. However, the distance of remote locations from large population centers restricts options for formal education and employment, leading to a disconnect with mainstream education and working systems and trends. Additionally, remote Aboriginal communities generally have poorer health and thus lower life expectancy, both because of limited healthcare access and, more recently, extreme weather events as a result of climate change.

One of the main methods of addressing the latter issue is adequate housing, which provides a direct buffer against the weather and allows for better adaptation to changing weather conditions. However, many indigenous peoples in Australia live in public rented housing, where tenancy regulations, poor housing quality, and seasonal overcrowding make it difficult to implement low-cost adaptation strategies, such as increasing shading, and taking advantage of electrical cooling appliances. Access to affordable, reliable electricity is also a strong factor in housing comfort and resident health. Despite these disadvantages, remote Aboriginal communities have proven to be resilient and enduring in the face of climate change, especially in ways that may not be evident to a Westernized, Eurocentric view of health; the indigenous view of health tends to be more holistic, taking into account not only physical well-being but being in touch with the land and sea surrounding the local community. At the same time, it is necessary to not mythicize these views and properly study whether communities are adapting properly to climate change, or maladapting while staying dependent on depleting resources and an unsustainable way of life. As such, this study measures adaptive capacity holistically in terms of information flow and social networks, acceptable and feasible adaptation options, local business and household viability, and a community's personal motivation for change.

The research is further informed by the concept of the Sustainable Livelihood Framework (SLF), which is commonly used in community development projects, but has been adapted to understand how climate adaptation can proceed in the two Aboriginal communities in this case study. SLF is capable of assessing what resources are available to improve adaptive capacity, and other conceptual frameworks have been added to take into consideration Aboriginal communities living on their traditional land and the effect of strong cultural values on the management of the local environment.

Through convergent triangulation, which entails the usage of multiple data sources such as semi-structured interviews and a focus group discussion, as well as quantitative data on the weather, a causal timeline was constructed to assess how changes within the last 60 years affected people's living arrangements and locations, daily behaviors, health, and community governance.

This research shows that many individuals are aware of both short-term weather fluctuations and long-term trends in the local climate, but this knowledge and other facets of natural knowledge are dwindling as individuals within these Aboriginal communities begin to lose touch with the countryside in which they reside. As a result, much of this local knowledge lives on only through individuals employed in outdoors jobs, such as park rangers. Ultimately, this trend indicates that the important traditional ecological knowledge (TEK) of the area has been deeply affected by changing weather patterns and access to land, despite local elders holding onto TEK and attempting to pass it on. In order to better preserve and sustain TEK, it is important to formalize channels through which TEK can be shared and documented, and included in NRM decisions. A successful example of this behavior occurs in the Aboriginal Ranger Group, which combines TEK with science and technology that are in line with the traditional cultural responsibility of "caring for country," or taking care of the land.

The study also concluded that a community's spatial reference, or its distance from major business centers such as towns, may have important implications for the community's adaptive capacity. Positively, the farther a community is from a large urban center, the more it can independently moderate government arrangements and take control in regulating its own resources and needs, which also lends to community-building and proactive, specialized problem-solving by local experts. One the other hand, these distances can cause locals to perceive outside support as distant and unresponsive, and make it difficult for outside experiences and ideas to filter into communities that could use this new knowledge to inform their NRM decisions. Additionally, interviewees confirmed the importance of housing as a mainstay in adapting to climate change, citing the use of electrical appliances and indoor socializing as methods of dealing with extreme weather conditions.

Overall, a community's independence contributes to higher social capital, measured in the strength of relationships among community members. However, this independence can impede adaptive capacity by limiting access to necessary resources, such as healthcare and education. Further, many traditional coping strategies for addressing extreme weather conditions are constantly compromised as the effects of climate change continue to manifest, making it imperative for public policy and government actions to properly support strong infrastructure, such as housing and electricity, in order to allow locals low-cost, long-term methods of climate adaptation. The role of blended knowledge, or the combination of TEK with contemporary scientific knowledge, is also important, as it maximizes the knowledge on both sides for the benefit of the community, and prevents outside government from implementing "cookie cutter" solutions that fail to address local complexities and nuances. Additionally, if promoted correctly, blended knowledge in the hands of locals empowers the community and encourages grassroots organizing, where the community itself will drive the changes and adaptations necessary for its continued survival.

All in all, effective adaptation must be both affordable and engaging for the community in order to be successful. Specific strategies for improving adaptive capacity include increasing understanding of social factors in the local community, making a holistic assessment of low-cost options for adaptation, and instituting governance structures and processes that empower local communities to make their own decisions. To achieve all of these goals, it is important to foster and center social capital in Aboriginal communities, rather than focus merely on physical assets. For Aboriginal communities in Australia particularly, existing development investments should be used as a foundation for integrating ideal physical and social outcomes by creating informational flow between remote communities and outside organizations, and by nurturing local TEK and cultural values.

Deedee Chao

The Effect of Policy Changes on Australian Community-Based Natural Resource Management Practices

Since the colonization of Australia, Western agricultural practices have dominated former indigenous practices of living with minimal impact and preserving the environment, leading to an unsustainable trajectory that the Australian government has been trying to amend via legislation. At the community level, one of the main approaches of addressing NRM has been Landcare, where local groups of volunteers carried out NRM activities and practices that were necessary in their communities. The collective success of roughly 4,000 Landcare groups throughout Australia has been attributed to their progressive attitudes and activities towards sustainability, egalitarian organization, collaboration with government, and focus on local decision-making. However, the introduction or revision of legislative measures and funding strategies has impacted Landcare, leading Cooke and Hemmings (2016) to examine how government policies on NRM affect the activities of Landcare groups that operate on the local level.

Cooke and Hemmings use the Habermasian theoretical lens to interpret their findings on the relationship between policy and Landcare groups. Habermas' theory focuses on the boundaries between systems and lifeworlds, the latter of which are intersubjective realms where people interact through shared cultural, social, and personal spaces. When a system asserts its power over a lifeworld and distorts it, Habermas describes it as colonization of the lifeworld. In this case, the government's Catchment Management Authorities (CMA), which were instituted to work with Landcare groups, and their actual effect and interactions with Landcare groups demonstrate the relationship between a system and a lifeworld, with failures in this relationship constituting a boundary crisis. Habermas' theory further states that social order can be viewed in one of two ways: a system in which people take on roles to fulfill the system's functions and goals, or a system that is held together by and exists because of the interpersonal relationships between people as people, rather than people as role-players. Both views must be taken into account in considering the boundary between a system and lifeworld, but come into constant tension when the system requires roles that are unnatural for the lifeworld. In other words, when governmental and economic systems pressure Landcare groups and their local communities to conform to external standards without regard for internal complexities, these local organizations and places face barriers to proactivity and effective NRM practices.

This framework was applied to case studies of four Landcare groups of up to 12 members each, all located within the Murray-Darling Basin, the catchment area of the Murray and Darling Rivers. This area constitutes 14% of the Australian mainland and is responsible for 41% of agricultural production, many practices of which are unsustainable. The data were collected from semi-structured interviews, historical documents from Landcare groups, governmental documents related to Landcare groups, and observations from Landcare meetings and field notes. These data were assembled via the textual analysis software NVivo and an intuitive analysis that created coded categories to compile themes and confirm intuitive analyses. Researchers aimed to create a chronology of policies and themes across the lifetime of each Landcare group in the study, while also studying the materials Landcare groups used to learn more about NRM.

From these categories, the study determined that a main issue was the dissonance between administrative timelines and natural timelines: funding and reporting schedules based on financial years, or program constraints of two years, made it difficult or impossible for five-year Landcare projects to receive initial or continued funding, as funding choices were based on yearly results that were impossible to achieve and incongruous with natural cycles. Administrative timelines and grassroots timelines also failed to match up, with arbitrary timelines for project proposals, seminars, and other requirements making it difficult for local Landcare volunteers to adjust their schedules to be able to attend. Further, these changes in administrative entities such as the CMAs and funding processes removed the control from local organizations in choosing which projects to carry out, making Landcare groups feel voiceless in influencing policy decisions. The lack of consideration for local inputs into policymaking processes also discouraged Landcare groups from communicating and providing feedback, exacerbating the feeling of powerlessness.

Another issue rested with the institutionalization of agricultural lives, where the government required compulsory certification courses on topics such as chemical handling, which were inaccessible because they were expensive and located faraway from where Landcare groups operated. The spatial problem persisted in an urban-versus-rural dichotomy as well, with urban groups being more comfortable with communicating with government officials than rural groups. Communication was also made difficult by a high turnover rate in CMA staff, who were meant to be the primary contacts for Landcare groups to get in touch with governmental agencies. Additionally, CMA coordinators were tasked with large workloads that involved bureaucratic tasks, relegating communication with Landcare groups to a much lower priority.

Overall, in the Habermasian framework, the demands of the system, or the government, failed to match up with the lifeworld needs of the Landcare groups, resulting in funding, communication, and trust issues.

Looking at the history of Landcare, the original structure that prioritized social relationships within local communities was replaced with the CMA and funding processes that prioritized the project proposals of individual farmers on individual properties, thus removing the emphasis on community that is imperative for local proactivity and successful community-based natural resource management practices. The boundary crisis between regional administrative entities and Landcare groups occurred because of colonization of local practices and relationships by external government agencies that failed to meet local needs. Colonization removed power from local communities by giving governmental CMAs the authority to dictate and approve local projects. However, since Landcare groups have been shown in the literature to be important players in sustainability practices, it is essential to figure out policies that will prevent Landcare groups from losing their vitality. Policies should reconcile regional or large-scale priorities with the priorities of local communities, and reconcile the vertical structure of governmental bureaucracy with the horizontal structure of Landcare groups. In the scheme of Habermas' system-lifeworld theory, this means that the system's demands should synchronize and re-prioritize the needs of the lifeworld in order not to impede and distort the lifeworld's processes and proper functioning.

The Role of Social Networks in Urban Forest Management in Hokkaido, Japan

In the 1960s and 70s, Japan experienced rapid economic growth that led to large-scale forest developments, which caused social anger and led to discussions surrounding nature conservation. Citizens' distrust in the government led to the National Forest creating a new communication system for the public to participate in forest management. Public participation has since become an important tool for the Natural Forest's management methods and planning processes, particularly for forests in urban areas where Japan's population is concentrated.

In general, public participation is considered a tool for decision-making that helps resolve conflicts of interest among different stakeholders. It also serves the normative goal of promoting democracy by giving the public a voice, and social learning by teaching the public about how to participate in governmental decision-making. Further, in order to facilitate public participation, organizations often have to develop new relationships, which in turn aid collaboration and mutual understanding. Collaboration fosters social capital, which is made up of reciprocal social networks. These social networks have been gaining more attention as a component of natural resource management, and Yamaki's (2016) paper examines the case study of the Nopporo Forest Regeneration Project (NFRP) in northern Japan through a social network analysis lens. The paper pinpoints all organizations formally and informally involved with the NFRP in forest management, compares how these two groups of organizations evaluate the current participatory approach being taken, and evaluates how well the current approach enhances collaborative social learning.

The case study focuses on the Nopporo National Forest, located in a suburb of Sapporo, the capital of Hokkaido, the northernmost island of Japan. Because of its urban location, the forest is used for many recreational activities, which must be balanced with conservation efforts. The NFRP was started in order to restore the forest after it was damaged by a typhoon in 2004. The National Forest organization recruited a committee of scholars, government organizations, and local stakeholders for the NFRP, as well as a number of citizen organizations to formally participate. The study used data collected from questionnaires sent to nine organizations formally related to the NFRP and 22 independent organizations that informally related to the NFRP as necessary. These questionnaires aimed to find out each organization's attitude to forest management, their evaluation of the current forest management approach, their evaluation of current collaboration with government agencies, and their evaluation of collaboration with other non-governmental organizations.

The study found that the National Forest organization was most central to the project, followed by the Hokkaido Prefectural Government, which was the regional government entity where the forest was located. On the whole, NFRP organizations had formalized, direct ties with actual governmental agencies through regular meetings and collaboration, whereas non-NFRP organizations had much more variation in their relationships with government agencies, usually approaching agencies only when necessary for their independent projects. Both NFRP and non-NFRP organizations agreed that it was important to share a common view on the direction of forest management and to pursue forest management with collaboration between government agencies and citizen organizations. However, there was a significant difference in the attitude towards the leadership of government agencies, suggesting that non-NFRP groups had a more negative view of government leadership than NFRP groups did. This difference could also be related to the independent nature of non-NFRP groups that do not rely on government leadership and diverge more often from the NFRP's official plans. In addition, non-NFRP organizations also evaluated the present state of forest management more negatively than did NFRP organizations. On average, non-NFRP organizations evaluated collaboration with the government negatively (below 3.0 on a 5-point Likert scale), and NFRP organizations evaluated collaboration at a higher average score.

From these results, the study concluded that the lack of a formal relationship structure between non-NFRP groups and government agencies resulted in fewer and weaker connections, thus creating only a minimal social network with minimal social capital, whereas NFRP groups benefited from formalized relationships with government agencies which provided regular, frequent meetings and collaborative opportunities. This highlights the importance of creating a formal, official social network of local public entities and government entities to facilitate relationship-building and trust-building in community-based natural resource management projects. Such a process spearheaded by the government also serves to increase public participation in other ways, such as public comments by individual citizens and relationships with other, less formally involved groups. This formalization becomes especially important in situations where public participation and collaboration would be difficult to achieve without external, motivating factors.

The Role of Multilevel Governance in Community-Based Environmental Management in Latin America

Governance of resource allocation usually occurs through one of three approaches, or a combination thereof: hierarchical approaches that use existing power structures, market-based approaches based on voluntary exchange, and community-based approaches that involve the cooperation of all parties involved. Because of the collaboration required by the last approach, community-based environmental management (CBEM) often requires multi-level governance (MLG), in which different actors work across horizontal and vertical dimensions administratively and jurisdictionally. It also involves a degree of decentralization of power and responsibility among the various parties. MLG has been promoted as a method that decentralizes and reassigns power to different levels and sectors of power, allowing local groups and non-governmental agents to play a larger role than they would under other methods of governance.

Sattler *et al.* (2016) focused on filling a research gap in CBEM methodology by investigating the role of MLG in successful CBEM scenarios, using case studies of four different CBEM cases in Latin America. The paper's goals are to first identify the individual actors and their societal spheres and governance levels, and then identify the roles that these actors play and how they interact with each other in promoting CBEM. The study collected primary data via personal interviews, observations of participants, and stakeholder discussions, as well as secondary data through reports, governmental documents, and websites.

In their analysis of individual actors, the researchers organized them by jurisdiction and sector, with the former split into subcategories of international, national, regional, local, and the latter into subcategories of civil society, market, state, and cross-sectoral. In the cross-comparison of these actors' roles, the study found that actors could be primarily active or passive in promoting CBEM. While state, market, and civil society actors were involved in both roles across the board, civil society actors tended to be extremely active while state actors were passive. In addition, actors involved at the local level were much more active than international actors, suggesting more proactivity the closer the actor was to the immediate community.

The interaction of actors across sectors was beneficial when each actor offered something essential that other actors could not; for example, civil society actors were best suited for initiating systemic change, while market actors were the best sources for professional services, and public actors were best for influencing legal changes. Across jurisdictions, MLG allowed for a better distribution of decision-making responsibilities, shifting power towards local groups who were most invested in successful CBEM. These interactions were also largely made possible by the proactivity of civil society actors, who took the lead in initiating CBEM. The non-profit, purpose-driven nature of civil society actors also allowed them to act as intermediaries to facilitate interaction between different parties. In these successful scenarios, it was also found that new actors, or institutions, were created to address a gap in community needs and further facilitate interaction between jurisdictions and sectors for better collaboration. Of course, in situations where various parties with different goals are working together, problems and conflicts arise, thus necessitating a forum for conflict resolution to be created in all cases as well.

In order to apply these findings to other communities, Sattler *et al.* (2016) determined that it would be necessary to focus on the context of each situation and identify the civil society actors who could take the lead on initiating CBEM strategies by empowering the community and creating communication channels between other actors. The researchers also recognize that long-term studies of successful CBEM in action would be beneficial to the field, as it would document the resilience and durability of CBEM.

Community-Based Rangeland Management in Mongolia Leads to Better Social Outcomes

Temperate grasslands are some of the ecosystems most threatened by climate change. With 80% of its territory consisting of this biome, Mongolia is particularly impacted. Recent droughts and tough winters, coupled with the transition from socialism to a still-undeveloped market economy, have led to the large-scale involvement of various external community-based rangeland management (CBRM) groups, who have sought to improve grassland conditions and the conditions of herders who work within these ecosystems.

CBNRM methods have been championed as a solution that both addresses biodiversity issues in local ecosystems and poverty issues in local communities. However, studies of CBNRM, and CBRM in particular, have shown mixed results. Fortunately, CBRM was employed at an almost national scale in this situation, creating the opportunity for a case study with the largest sample size in

CBRM research to date. Ulambayar *et al.* (2016) aims to delve into large-scale CBRM studies that will serve to better inform future CBNRM strategies.

The study focuses on analyzing the outcomes of CBRM through common pool resource (CPR) governance theory, which predicts that, in certain conditions and institutions, groups who share the same resources are capable of self-regulating resource usage. The researchers expected that post-socialist Mongolia lacked the institutions and enforced norms for proper rangeland governance, and that the CBRM groups would fill this gap and aid herders to better organize and govern pasture usage, leading to CBRM herders having better social and livelihood outcomes than non-CBRM herders. More specifically, the researchers expected to see stronger, wider social networks; more traditional and innovative management; higher incomes; and more assets in CBRM communities than non-CBRM communities.

The study took samples from four different ecological zones: mountain and forest steppe, steppe, eastern steppe, and desert steppe. Eighteen pairs of adjacent soums, or districts, were selected from each zone, with a CBRM and non-CBRM soum in each pair. From each soum, about five community groups were interviewed, with results collected from roughly five representative households per group. As a result, 706 households from 142 groups (65 non-CBRM and 77 CBRM), were surveyed. The study minimized pre-existing differences by analyzing poverty, leadership, and demographic indicators in each community. The independent variables were organization status (CBRM or non-CBRM) and ecological zone, and the dependent variables, or social outcomes, were livelihood status, social capital, and rangeland behaviors and practices. Each of these dependent variables are indices, with values calculated from all related survey items.

Statistical analyses of these variables' values showed that herders who were members of CBRM groups exhibited higher proactivity, more innovative and traditional CBRM methods, and more household assets than non-CBRM herders. However, the two groups did not have significant differences in social capital, income, or livestock number, and there was no correlation between ecological zone and organization status for any of the social outcomes. The latter implies that, in this case, CBRM led to similar social outcomes regardless of ecological location. CBRM member proactivity tended to manifest in three ways: the CBRM herders talked to experts about rangeland issues, joined local initiatives that aimed to improve resource usage, and initiated action to solve local problems. A comparative analysis of social outcomes in different ecological zones showed that groups in mountain and forest steppe had worse outcomes, with no significant difference among the other zones.

These results demonstrate that CBRM is associated with higher adaptive capacity through its correlation with proactive behaviors and strong traditional and innovative resource management methods, which are both social benefits for the steppe communities. However, these are all social outcomes, and conservation outcomes as a result of these resource management practices would need to be considered further to gauge the environmental effects of CBRM practices. The researchers attributed the similarity in social capital to Mongolian traditions of reciprocity among herder communities, which were already ingrained prior to CBRM intervention.

In addition, the researchers surmised that higher assets for CBRM members could be related to CBRM donor contributions of household assets, such as vehicles and fridges, and CBRM training that encouraged investment in technological assets. Since most CBRM groups were relatively young, the study predicts that there may be a bigger difference in social outcomes once CBRM teachings and practices have had more time to take hold. Further, the young market economy in Mongolia has not yet led to advanced price differentiation among low quality and high quality livestock; once the market becomes more developed, better quality livestock as the result of CBRM practices could also contribute to significant socioeconomic changes for CBRM herders. Markets are already proving to be a significant factor in social outcomes, as the mountain and forest steppe ecological zone, which is the furthest from markets, has the lowest social outcomes across the board, while the steppe herders closest to the markets have the highest incomes of all four zones.

The study also reminds us that most CBRM practices in the region have been catalyzed by external, non-Mongolian groups, and exportation of these practices or sustainability of these practices in the region should be internally developed by the government or further considered by involved NGOs and donor groups. Because of these early-stage positive outcomes, Ulambayar *et al.* (2016) suggest that more research be performed once CBRM practices have had a longer time to take effect in these Mongolian rangeland communities.

Using Prospective Structural Analysis as a Tool for Community-Based Natural Resource Management

In the face of extreme environmental challenges such as climate change and natural resource depletion, effectual natural resource management has become more important than ever. Recognizing that ecosystems and human societies coexist in social-ecological systems (SES) is an essential component of developing proper natural resource management methods, especially in the case of CBNRM strategies. CBNRM has been growing in popularity with donors and international organizations interested in natural resource management, as an alternative to governmental top-down approaches—however, in order to be efficient, CBNRM must invest in developing the capacity of local institutions and governance structures, as local communities often lack the organizational and financial infrastructure and resources necessary for adequate management. Some of the ways to strengthen CBNRM involve gaining more comprehensive understanding of local SES dynamics, implementing context-sensitive methodologies and tools, and using adaptable participatory tools that can encourage local involvement and empowerment.

Delgado-Serrano *et al.* (2016) investigate prospective structural analysis (PSA) as one such tool to accomplish all of the above. PSA is advantageous as it can structure descriptions of a system into the roles, both current and projected, of key variables and drivers within the system. This makes it possible to analyze the correlation between variables, which in turn allows actors to decide how to best proceed with each variable, taking into consideration its influence on other aspects of the system. To date, PSA and other strategic foresight methods have largely been used by the private sector and public institutions involved in large-scale projects, rather than CBNRM. Thus, Delgado-Serrano *et al.* (2016) also investigate the adaptations necessary to make PSA more applicable for CBNRM contexts.

PSA as a CBNRM tool was tested in three different SES in Colombia, Mexico, and Argentina, which were all facing different environmental challenges. All three were similar in that they dealt with managing common pool resources, and the local communities faced similar problems in their limited SES knowledge, communication issues and power dynamics with external actors, and problem of trade-offs between economic wellbeing and sustainability. At the same time, other differing factors provide a wide spectrum of scenarios and contexts in which PSA can be tested, allowing for a more comprehensive experiment. While PSA is usually conducted in three phases—listing the variables, describing their relationships, and analyzing the variables and their relationships—the researchers adapted it to CNBRM by adding two stages: an initial stage to select experts of the local SES, and a final stage to analyze and validate results with community members. The last stage was added with the intent of increasing local participation, a primary aim of CBNRM strategies. Additionally, technical PSA language was translated into more understandable layperson's terms.

Experts were selected via stakeholder mapping techniques that considered local inhabitants as internal stakeholders and outer organizations that had influence and understanding of SES as external stakeholders. In doing so, the researchers also realized that it would be more productive to separate the stakeholder groups while determining variables, in order to avoid power dynamics and ensure free speech from internal stakeholders. To determine variables, researchers presented a ready-made characterization of each SES to stakeholders and asked them to identify the most relevant variables from each; that is, the issues and matters most relevant to managing essential environmental challenges in the SES. Then, to analyze the relationships among chosen variables, a cross-impact analysis was performed, in which stakeholders were asked to rate a variable's influence from 0 to 3, with 0 being no influence. These data were input into MICMAC, an open-source software that calculated the direct influence and dependence of each variable, and recalculated the overall matrix of variables until a constant ranking emerged. Variables were grouped into five classifications: input variables that describe the system, characterized by strong influence and stability; stakes variables, characterized by high influence and dependency; regulator variables, characterized by moderate dependence and influence; autonomous variables, characterized by low influence; and output variables, which are easily affected by other variables and can predict the evolution of the SES. Finally, workshops were held to validate the graphed results of the software through consensus from a larger group of participants.

The results showed that in the Mexican SES, most variables were internal and linked to CBNRM activities in some way. In Colombia, a moderate number of variables were external, showing the influence of external organizations, but other variables demonstrated the need to build a stronger community identity. Similarly, in Argentina, external variables played a large role. The mapping of variables also showed participants and researchers where efforts should be concentrated and what should be taken into account when making decisions that would affect the SES; for example, in Colombia, there were few regulator variables but many stakes variables, suggesting that any decisions made would have a large ripple effect throughout the community and ecosystem, while in Mexico and Argentina, the opposite was true, with many regulators being potential methods of affecting one particular stake. Further, shifts in the graph with simulations of changed variables serve to demonstrate interdependence between variables: in Argentina, the market variable increased in dependency with shifts, suggesting that it could easily be influenced.

As a whole, the study determined that PSA proved to be a valuable tool in fulfilling certain CBNRM goals, such as understanding complexities of the local SES and the relationships between various components of the SES. Additionally, the PSA process promoted conversations between stakeholders and encouraged participants to reach consensus and shared ideas via the brainstorming and validation processes. For example, in the Colombian SES, the community collectively realized the benefits of ecotourism after going through the PSA process. Participation also increased the local community's knowledge of the SES and increased its capability for self-monitoring and self-governance of natural resources. However, Delgado-Serrano *et al.* also illuminated limitations of PSA as a CBNRM tool, such as its high degree of subjectivity in gathering all of its information and validation from stakeholders based on consensus. Overall, according to Delgado-Serrano *et al.,* in order to successfully utilize PSA for CBNRM purposes, it is imperative to properly select stakeholders who are experts on the SES, make participation accessible to laypeople, and take cultural dynamics into consideration. Then, PSA can start being employed as an effective CBNRM tool.

Conclusions

Even before climate change became an imminent issue, CBNRM had been recognized as a recommended approach for NRM in remote and rural communities that lacked access to more scientifically advanced, technological solutions to environmental and natural resource issues. Since the advent of climate change and its adverse effects on these communities—which have taken the form of extreme weather events that have harmed already scarce resources, livelihood practices, and public health—CBNRM has become increasingly prominent as a solution for vulnerable, remote areas across the world. As a result, scientific literature on the subject has increased, from monitoring the effects and natural responses of communities to climate change, to investigating the relationship between locals' adaptive strategies and governmental policy and non-profit intervention. Since the pool of knowledge is still relatively small, further data collection and analysis is necessary to better document the effects of CBNRM in addressing climate change. However, traditional precedents for community-based actions, as well as a recent large-scale study of CBNRM in agricultural practices, have shown promising results that encourage the continued usage and improvement of CBNRM methodologies. Continued research into both theoretical frameworks and practical tools for implementing CBNRM and expediting collaboration between external parties and local communities will be imperative for the future of CBNRM as a strategy for disadvantaged communities in the face of climate change and sustainability concerns.

References Cited

Cooke, P. R., Hemmings, B. C., 2016. Policy Change and Its Effect on Australian Community-based Natural Resource Management Practices. Journal of Education for Sustainable Development 10, 20–37.

Delgado-Serrano, M., Vanwildemeersch, P., London, S., Ortiz-Guerrero, C. E., Escalante Semerena, R., Rojas, M., 2016. Adapting prospective structural analysis to strengthen sustainable management and capacity building in community-based natural resource management contexts. Ecology and Society 21. doi:10.5751/ES-08505-210236

Jethi, R., Joshi, K., Chandra, N., 2016. Toward Climate Change and Community-Based Adaptation-Mitigation Strategies in Hill Agriculture. In: Bisht, J. K., Meena, V. S., Mishra, P. K., Pattanayak, A., editors. Conservation Agriculture: An Approach to Combat Climate Change in Indian Himalaya. Singapore: Springer Singapore. 185–202.

Pandit, A., Jain, A., Singha, R., Suting, A., Jamir, S., Pradhan, N. S., Choudhury, D., 2016. Community Perceptions and Responses to Climate Variability: Insights from the Himalayas. In: Salzmann, N., Huggel, C., Nussbaumer, S. U., Ziervogel, G., editors. Climate Change Adaptation Strategies – An Upstream-downstream Perspective. 179–194.

Race, D., Mathew, S., Campbell, M., Hampton, K., 2016. Understanding climate adaptation investments for communities living in desert Australia: experiences of indigenous communities. Climatic Change 139, 461–475.

Sattler, C., Schröter, B., Meyer, A., Giersch, G., Meyer, C., Matzdorf, B., 2016. Multilevel governance in community-based environmental management: a case study comparison from Latin America. Ecology and Society 21. doi:10.5751/ES-08475-210424

Ulambayar, T., Fernández-Giménez, M. E., Baival, B., Batjav, B., 2016. Social Outcomes of Community-based Rangeland Management in Mongolian Steppe Ecosystems. Conservation Letters. doi:10.1111/conl.12267

Yamaki, K., 2016. Role of social networks in urban forest management collaboration: A case study in northern Japan. Science Direct 18, 212–220.

Farming Practices and Climate Change

Grace Reckers

The increasing intensification of global climate change has illuminated a direct and interdependent relationship between changes in climate and farming practices around the world. Just as prominent production techniques used in agriculture and farming can emit large amounts of greenhouse gases (GHG), contributing to the greenhouse gas effects that lead to a rise in global temperatures, changes in climate adversely affect crop production and agricultural outcomes. With the recent commitments of nations across the world to reduce their carbon emissions by varying amounts over the next 30 years, environmental experts have consistently targeted the farming industry as one of the most important places to focus efforts. In places such as Brazil, for example, 18% of national GHG emissions come from livestock production alone, and 69% from deforestation to clear the way for commercial farming (Bogaerts *et al.* 2016). Numerous studies have proposed and assessed GHG mitigation techniques, such as improving ruminant diet, land sparing, reforestation to natural habitats, transition to no-tillage cultivation, and other practices that increase carbon sequestration that removes CO_2 from the atmosphere and creates carbon stocks in the soil. These stocks have been shown to not only reduce the growing atmospheric CO_2 levels, but soil with higher carbon content is healthier, more resilient, and can lead to greater crop production. Likewise, other studies have analyzed adaptation methods that aid crops in withstanding increases in global temperatures, such as altering planting dates, changing the varieties of seeds planted, developing drought-tolerant maize, and others. While there are many avenues to take to either mitigate emission rates of greenhouse gases or adapt to worsening climate conditions, there exists great potential for ever more effective strategies to reduce the carbon footprint of farming practices, all while improving agricultural productivity outcomes to feed the growing global population. The following studies explore some of the existing techniques and propose ways to implement successful strategies on a larger scale.

The Potential for Land Sparing to Offset Greenhouse Gas Emissions from Agriculture

The increase both in food demands across the world and in the amount of land dedicated to agricultural use has led to a 1% annual rise in measured global greenhouse gas (GHG) emissions. Under intensifying climate change, various countries have committed to reducing emissions, many targeting agricultural practices to do so. The UK has committed to an 80% reduction in GHG emissions by 2050 as compared to their 1990 levels. Land sparing strategies could be a viable method of mitigating emissions in agriculture. These strategies include improving efficiency to increase yields, cutting back land used for agriculture, and using this spared land from the cut back to restore natural habitats that would help sequester carbon.

Lamb *et al.* (2016) used the UK to estimate the potential for reducing GHG emissions through a number of land sparing strategies in agriculture and found a significant mitigation in emissions compared to 1990 levels through land sparing practices. The growing demands for food will necessitate greater agricultural production in 2050, whereas yields of commodities produced per hectare will likely increase as well through advances in technology and productivity.

The study began by predicting a range of feasible future yields using recorded data from patterns in agricultural yields along with projections of increases in food demands. Lamb *et al.* then made predictions of 2050 GHG emission rates from agricultural production practices that would change under land sparing, such as the manufacturing and transportation of chemicals and the energy use from the farm the equipment.

Next, Lamb *et al.* measured the projected 2050 emissions in the UK under land sparing strategies of increasing yields, which would allow for less farmland to be used, with the spared farmland then devoted to habitat restoration. Habitat restoration can mitigate GHG emissions through carbon sequestration of natural regeneration and from fewer fossil fuels used when growing bioenergy crops. This calculation was made using the emissions resulting from the production practices under the range of plausible future yields, from the decrease in land use, and from the six restoration of spared land. The study used six different restoration scenarios and calculated the projected mitigation of emissions from each for this calculation. These results were then compared with 1990 emission rates and assessed for mitigation potential.

Lamb *et al.* found that land-sparing practices demonstrate significant reduction in GHG emissions from 1990 levels, and at the upper bound of yield potential, the mitigation would exceed the commitment to an 80% reduction in GHG emissions. Five out of the six land change scenarios showed significant mitigation from projected emissions, with restoration from natural regeneration, broad-leaved forest, coniferous forest, *Miscanthus*, and short-rotation coppice resulting in significant mitigation of emissions and oilseed rape restoration having an insignificant effect on emission reduction. Finally, Lamb *et al.* looked at the feasibility of implementing land sparing strategies in the UK. Economic limitations were found to likely affect levels of mitigation. Changes in demand for food and prices of commodities will impact policies on reducing farmland usage, and even if increases in yield allow for less farmland to be used for agriculture, economic pressures will restrict the ability to restore the land spared to natural habitats. However, whether or not the spared land will be converted to natural regeneration or to bioenergy crops that would actively sequester carbon, Lamb *et al.* demonstrated that simply cutting back the amount of land used by improving agricultural efficiency has the potential to mitigate future GHG emissions in the

UK. The authors propose that given the scope of the mitigation potential and the ease of reducing land usage through improving yields, similar outcomes are likely to be observed in other agricultural regions of the world.

Models of Reforestation Productivity and Carbon Sequestration for Land Use and Climate Change Adaptation Planning in South Australia

Carbon sequestration has been recognized as a valuable agricultural method that benefits both the environment and the carbon market. Reforestation is one method to to increase carbon sequestration and can even be used as a way to generate carbon stocks that are then commoditized and used for purchase and sale in the carbon market. In order for those involved with this trade to optimize profits, investors use models of reforestation productivity that estimate the amounts of harvested carbon from specified plots of land to predict how much they will generate and sell. The level of productivity for carbon sequestration can be determined through measurements of biomass on a plot, which takes into account the amount of dry plant matter above the ground, and uses a root-to-shoot ratio to estimate the amount of biomass below ground. Amount of above ground biomass varies with plant arrangement, height, density, and volume. The calculated measure of total biomass is then multiplied by a factor of 0.496 to determine the tons of carbon sequestered. These differences in carbon stocks can then be compared to other plots with different climate, plant arrangement, and density averages.

Hobbs *et al.* (2016) designed models of carbon sequestration and reforestation productivity to estimate how a number of climate and planting scenarios would impact generated carbon stocks on reforested land. The study focused on South Australia and analyzed 264 reforested sites throughout the region with varying climate conditions and agricultural settings. The authors of the study were aware of the limitations and inaccuracies of pre-existing models, such as the FullCAM, which were known to unreliably estimate carbon stocks based on varying agricultural conditions. These limitations impact the carbon market, as some sellers are cheated of profits when predictions from models underestimate the amount of carbon generated by a given plot of land. Hobbs *et al.* recognized the need for more accurate ways to predict the amount of sequestered carbon and reforestation productivity to give reliable measurements based on a number of climate and planting scenarios, which would benefit the market and would allow planters to know which scenarios resulted in optimal levels of carbon productivity. In order to design more accurate models, Hobbs *et al.* focused on local measurements rather than the large-scale predictions made by FullCAM. These local measurements used in the study helped create models of carbon productivity that were more tailored towards community conditions.

The 264 reforested sites analyzed in the study were divided into two types of plant arrangements: woodlots of predominantly one species throughout the entire site, and environmental plantings of mixed species. These sites covered 10.2 million hectares across South Australia. Four climate conditions were applied as models for each site: the baseline climate conditions with historic temperatures, evaporation, and rainfall; mild warming and drying of an increase of 1°C, increase of 3% annual evaporation, and a decrease of 5% annual rainfall; moderate warming and drying of an increase of 2°C, increase of 6% evaporation, and decrease of 15% rainfall; and severe warming and drying with an increase of 3°C, increase of 8% evaporation, and decrease of 25% rainfall. Plant density was measured using spatial arrangements on each plot as well as data on plant volume, height, thickness, and weight. Hobbs *et al.* found average above ground biomass based on these measurements for smaller reforested plots and multiplied them by the area of larger sites. They then used GPS to assess how the four changing climate conditions and various plant arrangements affect total biomass and productivity rates across all 264 sites.

Hobbs *et al.* applied a forward step-wise model to demonstrate which reforested plant arrangements and climates optimized productivity. Changes in climate had the greatest significant effect on carbon sequestration, with rainfall having the biggest impact on productivity out of the three types of changes in climate. Varying the planting arrangements through spacing and density also both had significant effects on carbon productivity, although not as drastic as changes in climate. However, changes in climate do naturally change planting arrangements in many cases, and both are found to impact carbon sequestration. The authors propose that given these data on optimal climate conditions and planting arrangements for increasing reforestation productivity, people should be able to design their agricultural plots in such a way that they use reforestation to their advantage by sequestering and harvesting as much carbon as possible to then sell on the carbon market. They assert that their findings, although taken from South Australia, are applicable to other agricultural regions with similar climatic and ecological conditions as South Australia, and therefore their models should be used across the world to estimate carbon sequestration productivity on reforested land.

Climate Change Impacts on Soil Organic Carbon Stocks of Mediterranean Agricultural Areas: A Case Study in Northern Egypt

The risks of greenhouse gas (GHG) emissions can be mitigated through agricultural practices that actively sequester carbon, which removes CO_2 from the atmosphere. Sequestering carbon in soil not only reduces the amount that would otherwise be in the atmosphere, but creating carbon stocks in the soil is also helpful on farms to prevent erosion, improve soil fertility, and increase food production. These stocks of carbon held in the ground are called soil organic carbon (SOC). They have major impacts on the level of productivity in agriculture and are impacted by varying climate change scenarios. Changes in climate affect the ability of soil to sequester carbon, in turn influencing a plot of land's ability to produce food. Many other variables also affect rates of carbon sequestration, such as the land-use type characterized by what kinds of production happen on that soil, soil traits such as pH and bulk density, the erosion and drainage conditions of the site itself, and others. While many studies have looked at which factors affect the measure of SOC stocks in the topsoil, few have assessed the importance of SOC in the subsoil layers. Large amounts of SOC are found in depths lower than 25 cm, so there is a need to look at how these stocks in particular play a role in levels of production as influenced by various factors.

Muñoz-Rojas *et al.* (2017) studied how climate change impacts the soil's ability to sequester carbon at varying depths, durations of changes in climate, and uses of land. They also incorporated a number of other site and soil characteristics into their study to look at these trends in relation to SOC stocks. The authors were able to simulate atmospheric and climate change scenarios through Global Climate Models (GCMs) that create atmospheric circulation simulations and project the conditions of each change in climate. To measure the amount of carbon in the soil, Muñoz-Rojas *et al.* used CarboSOIL, which also analyzes how productive a site will be in sequestering carbon through different climates and land conditions. CarboSOIL quantifies the amount of SOC stocks at varying depths with different levels of precipitation, temperatures, site conditions, soil properties, and type of land being used. The study used agricultural areas in El Fayoum, Egypt because of it hot, arid, and Mediterranean climate that tends to have low SOC contents. Hot and dry regions have lower amounts of carbon in the soil because increases in temperature have been observed to inhibit the ability of soil to sequester carbon. This creates the potential to focus on these regions as places to increase rats of carbon sequestration to boost SOC stocks and improve agricultural productivity. Muñoz-Rojas *et al.* used 4 types of land-use scenarios in 28 different regions across El Fayoum.

The study recorded site and soil characteristics for all 4 land-use types at each of the 28 regions. These included annual precipitation, temperatures, elevation, slope, drainage, soil erosion, nitrogen in the soil, pH, cation exchange capacity, bulk density, field capacity, and land use/cover. It then projected the effects of changing climate on each site's ability to sequester carbon at varying depths, using 8 different Global Climate Models to create scenarios from predictions in changes in temperature and rainfall. The authors were able to validate the GCMs by comparing the measured and predicted SOC contents for each depth, which demonstrated accuracy as the R^2 values for these comparisons were 0.9715 for depths up to 50 cm and 0.9876 for depths of 50–75 cm. ANOVA was used to compare differences in SOC measures between types of land-use and different depths.

A sensitivity analysis and linear regression demonstrated causal relationships between changes in climate scenarios and SOC measurements. While increased temperatures significantly lowered SOC stocks across all land types at shallower depths (0–25 cm) as predicted, these increases in temperature actually led to higher SOC contents across all land-use types at the lower depths (50–75 cm). In terms of the land-use types themselves, olive groves had the lowest SOC measurements, likely due to the drought-tolerance of olive groves that means little irrigation is required. Dry regions with minimal irrigation tend to have lower SOC contents. Complex cultivation sites had the highest SOC measures. Muñoz-Rojas *et al.* recorded the impacts of each site and soil characteristic on SOC content at varying depths, and found some important causal relationships such as a site's drainage capacity and its ability to sequester carbon. Changes in climate had the most significant influence on carbon stocks.

Muñoz-Rojas *et al.* suggest that agricultural experts should focus on increasing SOC stocks to both improve crop production and to remove carbon dioxide from the atmosphere. The authors indicate the necessity of recommended management practices (RMPs) to incorporate the influence of each variable into the design of more productive farm practices that will optimize carbon sequestration. These RMPs need to be locally designed and site-specific rather than globally applied, as each region has vastly different site and soil characteristics as well as climate patterns. Water management will be an increasingly important aspect of these RMPs as levels of irrigation and drainage contribute to soil's ability to sequester carbon. These improved irrigation practices will also reduce the risks of drought. Overall, Muñoz-Rojas *et al.* predict that SOC contents in shallower depths will decrease in the short-term, but suggest that recommended management practices can help sequester carbon in lower depths that will eventually improve soil capacity to sequester carbon at all levels.

Evidence of Limited Carbon Sequestration in Soils Under No-Tillage Systems in the Cerrado of Brazil

Brazil has some of the highest rates of greenhouse gas emissions in the world, the majority of which come from deforestation and cattle grazing. The country has committed to reducing these rates by 2020. The Brazilian government created the "Action Plan for Low Carbon Agriculture" in 2010 to propose and implement new mitigation practices to reduce their carbon footprint. One prominent initiative created by the "Action Plan" aims to convert land that has been used for conventional tillage (CT) practices to no-tillage (NT) practices, with the goal of increasing the amount of NT land from 32 to 40 million ha by 2020. The bulk of this land lies in the Cerrado region of Brazil, a savannah covering 23% of the country.

Non-tillage land can benefit outcomes of farming practices in numerous ways. Not only does it reduce soil erosion and decrease the need for labor, fuel, and machinery, but non-tillage crop systems have also been found to increase soil organic carbon (SOC) stocks in agriculture. This is important because SOC levels indicate success in mitigating CO_2 emissions, the most common greenhouse gas emitted through agriculture. Land with high SOC levels also tend to be healthier, more resilient, less prone to erosion, and facilitate a stronger diversity of necessary microbes. Conventional tillage land tends to have higher rates of erosion from the destructive practices and have a decreased ability to sequester carbon. Conventional tillage also decreases the amount of soil in a plot of land by 17% after 26 years of using the practice, which lowers SOC concentrations and increases risk of erosion. These are the reasons for which initiatives have aimed to convert CT land to NT crop systems throughout Brazil.

Corbeels *et al.* (2016) studied the increase in soil organic carbon stocks from the transition of conventional tillage to no-tillage crop systems for maize and soybean farms in the Cerrado region of Brazil. They acquired data from the Rio Verde municipality in the

Goiás state, taking measurements from just over 5,000 ha of land. The study compared levels of carbon present at five varying depths (0–5 cm, 5–10 cm, 10–20 cm, 20–30 cm, and 30–40 cm) across four different types of land uses: natural Cerrado, pasture, conventional tillage, and no-tillage. The NT plots were further divided by the time passed since they transitioned from conventional to no-tillage, beginning at one year and ending at twenty-one years of practices without tillage (indicated as NT-1 being the youngest plot to transition, having transitioned from CT to NT just one year before the start of the study, then NT-2, up to NT-21). Measurements for the study began in 2003, so NT-1 was the plot of land that transitioned to NT in 2002, while NT-11 had been NT since 1992. Corbeels *et al.* then used stainless steel cylinders to collect and analyze soil samples at each depth for every plot. They based calculations on the bulk density from the soil samples, using estimations of the SOC stocks for each density from equal soil mass-depth data. Corbeels *et al.* carried out this soil analysis procedure in both 2003 and 2011, and applied a linear mixed model to assess the impact of site on SOC concentrations for both years.

Prior studies demonstrated that SOC concentrations were greater in NT crop systems and lower in plots with CT techniques, which was reflected in the results of the study. Data from Corbeels *et al.* showed that natural Cerrado land had the highest SOC stocks, then NT (averaged across NT-1 through NT-21), then pastures, and finally CT crop systems. The study found that the greatest SOC stocks were found within the top layer of soil for all land uses, and while carbon levels had an overall decrease with depth for each plot, the stratification of carbon levels differed between land use types. Surprisingly, the study reflected significantly ($P<0.05$) lesser carbon in NT-1 than in CT samples in 2003. Since NT-1 had just recently transitioned to no-tillage, Corbeels *et al.* proposed that this result may be due to the large disturbance to the soil that occurred during the transition, which would have decreased rates of carbon sequestration. It is likely that a large amount of soil eroded during the transition, which would have impacted SOC stocks. There was not much observed change in SOC levels between 2003 and 2011 across the natural Cerrado plots, which was to be expected as the crop system did not change.

Corbeels *et al.* also found that it takes about 11–14 years for SOC stocks to return to their pre-CT levels through NT practices, at which point the amount of carbon present in soil levels off and little carbon can be added. This should not be a disincentive to the transition from CT to NT, however. The study also looked at the holistic benefits of NT crop systems, such as better soil fertility, less required labor and materials, greater ecosystem resilience, increased diversity of beneficial microbes, reduced vulnerability to erosion, and others. Ultimately, in addition to the mitigation of CO_2 through higher levels of SOC stocks, Corbeels *et al.* found that NT practices could be a smart economic decision when replacing CT techniques.

Comparing Measurements of GHG Emissions Between Conventional Brazilian Farms and Those with Sustainability Programs

Greenhouse gasses (GHG) constitute a number of gasses (CO_2 being the most prevalent) that are released from the earth's surface and trap heat in the atmosphere. They have become of primary interest to many environmentalists because of their impacts on agriculture, human health, ecology, and other environmental systems. Countries across the world have committed to reducing GHG emissions due to general increased recognition of their detrimental effects. One such country, Brazil, aims for a 37% reduction of their 2005 emission values by 2025. As the second-largest producer of beef in the world, Brazil has acknowledged the notable fraction of GHG emissions derived from livestock production (18% of Brazil's annual GHG) and the particular relation between the effects of cattle ranching and beef production on national emissions.

Agricultural experts have identified a number of different farming techniques that could potentially reduce GHG emissions through a process known as intensification. The practices involved in intensification are meant to make beef production more efficient both by increasing the amount of beef produced as well as by decreasing GHG emissions measured from the cattle industry each year. These emissions come from a number of different sources related to beef production, whether derived from the cows themselves or from the products and conditions necessary for the cattle industry.

The ultimate aim of intensification is to be able to stock more cows per hectare of farmland and to reduce the slaughter age. This is achieved through improving pasture management by rotational grazing and addition of nutrients to the soil, and by increasing the productive output of cattle through supplements and feedlots. These two intensification practices, used in tandem, have the potential to produce more beef per farm and to reduce overall emissions.

Bogaerts *et al.* (2016) surveyed 40 cattle farms across five northern municipalities of Brazil, 21 of which operated farms without sustainability programs, and 19 of which operated farms with sustainability programs. Four of these sustainability programs were identified as such due to their ability to work closely with cattle farmers to develop environmentally considerate management practices, and one of these programs was identified from its certification rewarding sustainable management. Each of the sustainable programs identified in the study aimed to improve cattle productivity by lowering the slaughter age and by increasing the stocking rates of cattle, and to promote better pasture management through pasture rehabilitation and rotational grazing.

In order to control for other characteristics that differed between farms, Bogaerts *et al.* used qualitative assessments of the geographic proximity, kind of operation, and operation size at each program and non-program site and ensured comparability in these characteristics. Emission rates were then measured for each program and non-program farm using the Cool Farm Tool to calculate GHG that derived from beef production processes and from the cattle themselves. This tool measured GHG that came from manure deposits, fertilizer manufacturing, pesticide manufacturing, feed production, enteric fermentation (cattle CH_4 emissions), and N_2O from fertilizer left on soil. These rates were then categorized into two sources of emissions: those directly from beef production, calculated into a measurement of emissions per kilogram of beef production (kg of CO_2e/kg of beef produced); and those from contributions of the land, calculated into a measurement of emissions per hectare of pasture (kg of CO_2e/ha/yr). The rates of emissions from program and non-program farms were compared for both categories once farm size, location, number of cows, years farm had been owned, and area of pasture had been controlled for through a series of linear regressions. These other explanatory factors that may otherwise have played a part in differences in emissions either per kg of beef produced or per hectare of pasture were demonstrated not to be statistically significant, meaning that each of the 40 identified farms were comparable in terms of farm characteristics.

Analysis of GHG emission rates per kilogram of beef produced in Bogaerts *et al.* demonstrated a 18.6% reduction in emissions per kg of beef produced on program farms when compared to that of non-program farms. This came from the 44.7 (±21.4) kg of CO_2e/kg of beef produced on non-program farms measured against the 36.4 (±14.6) kg of CO_2e/kg of beef produced on program farms. The difference in emission rates was not statistically significant.

Bogaerts *et al.* then compared GHG emissions per hectare of pasture between program and non-program farms and found that program farms had on average 4,552.2 (±2,106.6) kg of CO_2e/ha/yr, while non-program farms had on average 4,483.5 (±3,397.2) kg of CO_2e/ha/yr. This indicated a 67.8 kg of CO_2e/ha/yr difference in emissions per hectare of pasture between program and non-program farms, which was not statistically significant, and was presumably due to the greater animals per hectare found on program farms (2.25 (±0.9) animals per hectare) compared to that of non-program farms (1.92 (±1.4) animals per hectare).

Finally, total GHG emissions per year were calculated from adding up all sources and measurements of emissions for each category of farm. For farms with sustainability programs, the median total emissions per year were 2,081.6 tCO_2e. The total emissions for farms without programs were 2,512.4 tCO_2e per year. This took into account emissions from enteric fermentation, manure, fertilizers, pesticides, and feed. Overall, there was a small reduction in GHG emissions for farms that participated in sustainability programs, although this reduction was not statistically significant at each measure of source of emission. While the initial data on emissions per hectare indicated that program farms actually had greater emissions than per hectare GHG on non-program farms, this did not take into account the improvements in pasture management and productivity that favored program farms and were measured in the per-kg emissions that were smaller for program farms.

Evidence of increased productivity and reduced emissions on Brazilian farms participating in sustainability programs demonstrates a transition from conventional Brazilian open pasture farming to the new practices of intensification, even when the differences between emission rates were not statistically significant. This becomes more compelling when looking at the increased herd size of program farms after participating in the sustainability programs for multiple years in a row, where stocking rates increased by 23% for program farms and slaughter age decreased by 3.4 months. This allows farmers to practice more efficient beef production while potentially reduction their GHG emissions. However, this comes at the cost of a loss of conventional farming practices. No longer are Brazilian farmers encouraged to raise cattle on open pastures; but there is now an emphasis on cramming in as many cows as productively possible on smaller plots of land, and on settling on younger ages at which to slaughter the cattle for beef production. The benefit of using less land, however, is not only that it emits fewer emissions due to production, but also that it leads to less deforestation, an issue of growing concern whose relation to GHG emissions has been studied but not discussed in the Bogaerts *et al.* study.

Integrating Climate Change Mitigation and Adaptation in Agriculture and Forestry: Opportunities and Trade-Offs

Mitigation and adaptation are two prominent strategies used in the agricultural industry in response to worsening climate change. Many countries have recognized the need for national policies to improve mitigation, or the reduction of greenhouse gas emissions generated by agriculture. Others have focused on the importance of adaptation strategies that help human and plant populations better adjust to changes in the climate so that they are not as vulnerable to its damaging effects. These two approaches are typically treated as distinct from one another. While there is growing awareness of the dangers of climate change, and studies have looked at the success of mitigation and adaptation separately, there has been little research or documentation of the impacts of combining mitigation and adaptation strategies.

Locatelli *et al.* (2016) analyzed various landscape management practices and recorded the observed outcomes of incorporating both the mitigation and adaptation techniques. Their observations demonstrated occasional trade-offs in combining the approaches. The study found that while certain landscape management practices that are targeted to mitigate greenhouse gas emissions may in fact reduce those emissions, those practices can have the reverse effect on adaptation, making ecosystems more vulnerable to the harms of climate change. Likewise, agricultural practices that improve plant adaptation can in turn reduce their ability to sequester carbon and result in an increase in greenhouse gas emissions. The study categorized the observed outcomes of the various landscape management practices into three groups: practices that benefit adaptation while damaging the effects of mitigation, those that benefit mitigation while damaging the effects of adaptation, and those in which both strategies benefit, meaning the outcomes helped reduce greenhouse gas emissions while making ecosystems better adapted and less vulnerable to the impacts of climate change.

Locatelli *et al.* used optimization analysis to determine which practices resulted in the greatest mitigation of emissions and the most effective adaptation to changes in climate. In comparing 274 existing landscape management practices and documenting their impacts on adaptation and mitigation, they were able to narrow down which were most effective strategies. Generally, ecosystem and forestry conservation and soil management practices had the optimal outcomes for both adaptation and mitigation. The restoration of sustainable ecosystems through improved soil management in particular increased carbon sequestration, hence reduced emissions, and made the soil more resilient to climate change, meaning the ecosystem was better adapted and less vulnerable. The study found that by using conservation techniques that sequestered carbon in stocks beneath the soil, the soil had increased productive outputs, facilitated a greater diversity of plants and microorganisms, and resulted in stronger ecosystems as a whole. Locatelli *et al.* noted that this increase in diversity of flora would potentially lead to improved health outcomes of surrounding human populations as well.

One of the most prominent trade-offs observed out of all 274 combined landscape management practices was when farmers shorten the planting rotation schedule to better adapt the crops, which then reduces carbon stocks in the soil. Another was when attempting to increase carbon sequestration, farmers grow monocultures, which then impedes resilience and makes the entire ecosystem more susceptible to the adverse effects of climate change.

Once the optimal landscape management practices were determined from combined mitigation and adaptation strategies, Locatelli *et al.* looked at the likelihood of various communities adopting these practices. The study used surveys and socioeconomic trends to conclude that there was popular support from stakeholders and investors in favor of combined adaptation and mitigation practices due to cost-saving benefits and greater resilience of crops, but that high costs would limit support. Affordable practices would be necessary to implement the improvements on a large scale. Another risk is of increased institutional complexity when it came to actually applying the practices to various regions across the world, as differing systems of governance and agricultural regulations could interfere with efficiency. Locatelli *et al.* concluded that for these optimal strategies to be successful, investors will need to work directly with the local communities to assess their needs and desires, and plan accordingly when it comes to which practices are most appropriate and will result in the greatest rates of adherence.

Assessing Adoption and Adherence Rates of Drought Tolerant Maize in Sub-Saharan Africa

Maize is one of the most important crops in sub-Saharan Africa (SSA) because of its protein, vitamin, and mineral content that make it a staple for millions of people. However, it is also one of the most susceptible crops to the detrimental effects of widespread drought because of its dependence on rainfall when irrigation infrastructure is not in place. Climate change has increased the severity of droughts throughout the African continent and has made them less predictable, worsening their effects. Climate change models predict a 2°C rise in mean annual temperatures across sub-Saharan Africa between the late twentieth and the mid-twenty first centuries. Twenty-five percent of maize grown in SSA is already affected by frequent droughts, which can reduce crop outputs by 10–25%. Due to the intensified threats of climate change and its continued impacts on such a staple crop, seed companies have targeted maize as a crop to focus on for drought tolerant strain development.

Drought Tolerant Maize for Africa (DTMA), an initiative sponsored by both public and private investments, develops and distributes drought tolerant (DT) maize throughout SSA countries. The initiative created 160 types of DT maize rom 2007-2013, testing the DT varieties through multi-location on-farm techniques. The results of these tests consistently found higher crop yields from the DT maize when compared to commercial breeds. However, numerous studies have also found that despite national promotion of DT maize, rates of adoption and adherence by farmers throughout many SSA countries have been low. The DTMA seeks causes for these low adoption rates, especially in the face of such devastating effects of climate change that intensify the drought.

Fisher *et al.* (2016) evaluates these causes and proposes methods of resolving the reasons behind the lack of widespread farmer use of DT maize. The study looks both at the adoption rates as well as the determinants of low adherence, analyzing three countries in eastern SSA (Ethiopia, Tanzania, and Uganda) and three from southern SSA (Malawi, Zambia, and Zimbabwe). The purpose was to identify strategies that would increase adoption and adherence of DT maize amongst farmers in each of these countries. Part of Fisher *et al.*'s hypothesis was that farmers' perceptions of climate change are some of the primary determinants of low adoption rates of DT maize. Using a five-page farmer questionnaire in 2013, Fisher *et al.* found that the majority of participating farmers observed changes in climate conditions over the previous five years, such as 86% of those in Ethiopia reporting increases in temperature and 79% of those in Uganda noticing a decline in rainfall. Responses to this questionnaire were gathered from each of the six countries in maize-growing regions. They were aggregated from six sample regions, drawing 400 to 900 household responses from each sample. Farmers in each country reported longer-lasting droughts compared to five years ago and discussed the detrimental effects of these droughts on crop yield. This confirmed the assumption that farmers did observe changes in climate, which would presumably increase their interest in drought tolerant varieties to combat variable climate effects.

Farmers were then asked about a number of household characteristics, such as the average age of household heads, labor availability, educational levels, cultivated area, livestock units, maize self-sufficiency, and likelihood of accessing credit to purchase maize seeds. These statistics were gathered for each sample size and compiled by country. Information on the percentages of each sample region already growing DT versus those growing non-DT maize was also included. Fisher *et al.* then used the household statistics to compare adoption rates across different demographics, finding younger household heads to be more likely to grow DT maize compared to older household heads, and farmers with higher levels of education to be more likely than farmers with fewer years in school. Those who had greater access to credit to purchase seeds had higher rates of DT maize-usage, although this association was not statistically significant. However, there was a statistically significant difference in use of DT maize between smaller farms and larger farms, with higher usage rates for the latter, suggesting that those with greater profits from more land to cultivate had increased access to DT maize. This corresponded with results demonstrating that some of the greatest determinants of low adoption rates were the lack of resources reported by farmers, such as constraints on cash, labor, and land. The high cost of DT seeds was another one of the greatest determinants according to reports gathered from farmers in the study. The final major determinant had to do with the amount and type of information households received about DT varieties, with those receiving promotional information about DT maize being significantly more likely to grow it themselves.

Finally, Fisher *et al.* used the statistical findings to propose policies and other methods to increase adoption rates across SSA. Malawi had the greatest percentage of farmers already growing DT maize at 61%, while the rest of the countries only found 9–25% of farmers planting DT maize. This is likely due to the Farm Input Subsidy Program (FISP) in place in Malawi that lowers the cost of fertilizer and seeds for maize farmers. The impacts of this program on the high adoption of DT maize implied the success of subsidy programs in increasing DT usage. Fisher *et al.* concluded that increasing the percentage of farmers who plant DT maize would primarily depend on reducing costs, disseminating more promotional materials about the benefits of DT varieties, and removing travel barriers to markets where DT is sold.

Impact of Progressive Global Warming on the Global-Scale Yield of Maize and Soybean

The impacts of climate change on plants are primarily due to increasing global temperatures that affect the length of growing seasons and planting dates. These changes are generally understood to reduce crop yields, although some regions report that the warming has positive effects. Since maize and soybean are the two most important crops globally, with maize serving as the primary staple for over 900 million people across the world and soybean as a rapidly growing source of calories and protein, efforts have been made to understand climate change's impacts on these two crops in particular. Two adaptation strategies have been identified as the most effective methods for counteracting the effects of global warming for maize and soybean crops, the first being to alter the planting date. Increasing temperatures can shorten the growing season by inhibiting physiological plant development at later stages, so planting earlier in the year when it's cooler can allow for a longer growing season. The second adaptation strategy is to change the variety of the crop. Some varieties produce more under warmer temperatures, while others have optimized yields under cooler ones. Various studies have found that both of these adaptations have been found to significantly reduce the negative effects of global warming on crop yield.

Rose *et al.* (2016) used climate change modeling to assess the impacts of incremental increases in global temperatures on maize and soybean yields, both on national and global scales. The study used the General Large Area Model (GLAM) as a simulation model to measure the effects of climate change on crop yield in the ten countries with the highest production of both maize and soybean. Three different varieties of each crop were run through the GLAM simulation, and two GLAM simulations were run for each variety to account for rain-fed versus fully-irrigated farms. The effects of climate change on crop yields were assessed using records from 1961–1990 as a baseline to measure temperature increases of 0.5°C to 4°C. This was done through interpolating the monthly temperatures, rainfall, and cloud cover for each country over the thirty-year period and using a pattern-scaling approach to determine the future climates of each location under the temperatures of 0.5°C, 1.0°C, 1.5°C, 2.0°C, 2.5°C, 3.0°C, and 4.0°C with monthly averages from 1961 to 1990 as the baseline for each incremental temperature. Finally, the study included both adaptation strategies in the GLAM and reported their effects on yield for each incremental temperature by aggregating results of changing crop variety and planting date with and without adaptations.

Rose *et al.* found that each temperature increase of 0.5°C from 0.5°C to 4.0°C caused an overall global reduction in yield for both crops compared with baseline 1961–1990 data. Maize experienced a greater negative impact of global warming on yield than did soybean. For both crops, adaptation strategies helped limit the negative effects of warming globally, while some countries experienced less of an impact from adaptation strategies than did others. Looking at the effects of incremental temperature increases on crop yield for each country demonstrated differences in impacts for simulations both with and without adaptations. Each country experienced at least a 30% reduction in yield for the 4.0°C temperature increase, while adaptations resulted in a much smaller reduction for most countries. The only country to see no significant difference in yield when adaptation strategies were in place was Paraguay, where neither altering the planting date nor changing the crop variety had an effect on yield either crop. When adaptation strategies were not in place, the only country to see an increase in yields under warmer temperatures was India, and that was observed when temperatures exceeded 3.0°C.

The effects of implementing adaptation strategies were greater for soybeans than for maize. For maize, yields saw a 10–20% reduction when increasing temperatures from 0.5°C to 1.0°C alone without adaptations. However, with adaptation, six out of the ten countries saw no effect on yield when increasing temperatures by 0.5°C, while increases above that resulted in reductions. For soybeans, adaptation strategies actually increased yield for China, India, and Korea, and five other countries saw smaller reductions of soybean yield when adaptations were in place. These affects were greater at the lower end of the temperature intervals, meaning adaptation was more effective in preventing yield reduction for lesser increases in temperature, but adaptation strategies were overall effective for maize and soybean production up to 4.0°C.

Rose *et al.* concluded that while a few countries saw increased crop yield under specific increased temperatures while adaptation strategies were in place, the overall results of the climate change GLAM on yields of soybeans and maize throughout all ten countries indicate that warming temperatures have increasingly negative effects on crop yields. Adaptation strategies will have varying effects on each country, and while some countries may not experience much of a change in reduction when adaptation is implemented, most countries experienced smaller yield reductions with both adaptation strategies, and some even increased yields when temperatures rose.

Conclusions

The great variety of available methods that could help the farming industry reduce greenhouse gas emissions and adapt to changes in climate gives hope for further use and development of sustainable practices. While many of these strategies are already well known and trusted by farmers across the world, challenges remain in implementing others that may be far too costly or otherwise not feasible. Economic constraints have become limiting factors for farmers and local governments who recognize the need for changes in agricultural practices, but who do not have the material means to do so. Political beliefs have arisen as another obstacle, as battles over resource rights and land usage continue to hinder the adoption of necessary strategies that could actually lower costs and mitigate GHG emissions. Farmers tend not to make the claim that climate change is a hoax, as they are intimately familiar with how recent rises in temperatures have impacted crop production, but they often have to deal with bureaucratic authorities who do not prioritize sustainable practices and who make it challenging to acquire the funds or rights to implement necessary changes. As several of the reviews discussed, it is crucial to target these mitigation and adaptation strategies to local communities rather than to apply them on a global scale, as methods that may be effective in one region may have adverse effects in another. Similarly, cultural conventions in one society may align with particular farming practices, while another society may reject them for legitimate reasons that need to be respected. Finally, there needs to be a greater global recognition on the immediacy of the climate change issue and governments must be on board with prioritizing intervention strategies. A coordinated effort to combat the increasingly damaging effects has become necessary, and as agriculture and livestock production are responsible for some of the greatest rates of greenhouse gas emissions, a greater focus should be on improving and implementing sustainable practices within farming industries across the globe.

References Cited

Bogaerts, M., Cirhigiri, L., Robinson, I., Rodkin, M., Hajjar, R., Costa, C., Newton, P., 2016. Climate change mitigation through intensified pasture management: Estimating greenhouse gas emissions on cattle farms in the Brazilian Amazon. CGIAR 188.

Corbeels, M., Marchão, R.L., Neto, M.S., Ferreir, E.G., Madari, B.E., Sopel, E., Brito, O.R., 2016. Evidence of limited carbon sequestration in soils under no-tillage systems in the Cerrado of Brazil. Nature 6, 1-8.

Fisher, M., Abate, T., Lunduka, R., Asnake, W., Alemayehu, Y., Madulu, R., 2016. Drought tolerant maize for farmer adaptation to drought in sub-Saharan Africa: Determinants of adoption in eastern and southern Africa. Climatic Change 133, 283-299.

Hobbs, T., Neumann, C., Meyer, W., Moon, T., Bryan, B. 2016. Models of reforestation productivity and carbon sequestration for land use and climate change adaptation planning in South Australia. Journal of Environmental Management 181, 279-288.

Lamb, A., Green, R., Bateman, I., Broadmeadow, M., Bruce, T., Burney, J., Carey, P., Chadwick, D., Crane, E., Field, R., Goulding, K., Griffiths, H., Hastings, A., Kasoar, T., Kindred, D., Phalan, B., Pickett, J., Smith, P., Wall, E., Ermgassen, E., Balmford, A., 2016. The potential for land sparing to offset greenhouse gas emissions from agriculture. Nature Climate Change 6, 488-492.

Locatelli, B., Pavageau, C., Pramova, E., Di Gregorio, M. 2016. Integrating climate change mitigation and adaptation in agriculture and forestry: opportunities and trade-offs. WIRE's Climate Change 6, 585-598.

Muñoz-Rojas, M., Abd-Elmabod, S., Zavala, L., de la Rosa, D., Jordán, A. 2017. Climate change impacts on soil organic carbon stocks of Mediterranean agricultural areas: A case study in Northern Egypt. Science Direct 238, 142-152.

Rose, G., Osborne, T., Greatrex, H., Wheeler, T., 2016. Impact of progressive global warming on the global-scale yield of maize and soybean. Climatic Change 134, 417–428.

Contribution of Food Production and Consumption to Climate Change

Sarah Whitney

During my semester abroad, the relation between food production and climate change became extremely apparent and integral to my studies as an environmental analysis major. Fall semester of 2017 I studied at the Danish International School in Copenhagen, Denmark. My core course was called Sustainable Development in Northern Europe, and it illuminated the various avenues that the Danish government has taken in implementing national environmentally sustainable practices.

Despite extremely progressive environmental energy policies, practices, and infrastructure, Denmark was ranked the fourth most unsustainable nation in the world according to the 2014 WWF's Living Planet Report (WWF, 2014). This higher-than-expected rating was heavily attributed to the Danish production of pork. Denmark has almost 5,000 pig farms, and produces approximately 28 million pigs, from which approximately 90% of pork production is exported (Danish Pig, 2017). Furthermore, a major portion of the feed is imported from South America. This statistic sparked my interest in the topic of agriculture and livestock, and their contribution to climate change. My curiosity ventured from people's opinions and knowledge of agriculture and climate change, to specific statistics of food production's contribution to climate change, meat alternatives, and possible mitigation strategies revolving around food and agriculture.

My research demonstrated that food is a very personal subject. Cultures, traditions, connections, and even personal identities revolve around food. Thus, there was a strong unwillingness to reduce meat consumption or change eating habits. Even though many individuals were unaware of their food's contribution to climate change, once informed they felt that their personal contribution was insignificant. The study by Macdiarmid *et al.* (2016) illuminated the importance of food in our lives and a deeper resistance of change. Further research gave statistical significance to the fact that meat consumption is important to one's personal identity. A change in beef consumption specifically will not occur unless one's food consumption is a threat to their health or socially unacceptable. Such a change in perception must be as easy and accessible as possible if it is to occur. As it turns out, people would rather reduce proportion sizes than try other forms of meat. Furthermore, most respondents viewed switching to a vegetarian option as highly unfavorable.

Taking a more analytical approach to food production, I researched beef consumption specifically. Multi-functionality of cows that produce dairy products and meat is statistically more sustainable. Overall dairy-producing cows require the use of more fossil fuels and water than grass-fed beef. Nevertheless, grass-fed beef has greater global warming potential than dairy cows. A major part of the environmental impact of beef production, as well as of other meat production, is the production of soy or grain feed. Matassa *et al.* (2015) give alternatives such as fungi or bacterial biomass feed. Microbial production for feed has an immensely lower environmental footprint, which can reduce the total environmental degradation of meat production.

My research then led me to exploring policy options regarding agricultural and climate change. Methods of mitigation include restoring soil quality to increase crop production efficiency, using soils as carbon sinks, reducing the use of fertilizers, and containing and managing animal manure and their emissions. Yang *et al.* (2017) warn about the use of ALCA, a new form of life cycle assessment, in policy making. They state that it gives a warped perception of sustainability and reduces the significance of local food consumption and thus should not be used in policy decisions. A carbon tax on meat products is suggested to incentivize a change in meat consumption patterns due to the success of a case study in Spain. Additional investigation demonstrated that people might respond positively to such a tax. Today, consumers are invested in reducing their carbon footprint, and will pay more for environmentally friendly food products. The variance and complexity of my full examination is recorded in the summaries below.

Public Perception and Opinion of Climate Change and Meat Consumption

Macdiarmid *et al.* (2016) seek to comprehend the level of public awareness of the negative environmental impact of the meat industry, and the willingness to adopt dietary changes. Their study indicates that meat consumption is more than just a way of eating, resonating with cultural, personal, and social sectors of our lives. The study was structured as four individual interviews and twelve focus groups comprising adult participants from all different socio-economic backgrounds from various regions in Scotland. The collected data were organized by theme and analyzed. Overall trends included a general lack of understanding of the connection between meat consumption and climate change, reduction of individual contribution to reducing the effects of climate change due to personal dietary choices, and an unwillingness to reduce meat consumption. The act of eating meat was seen as a socially, emotionally, and physically pleasurable experience. Meat consumption was associated with traditions, personal, and family values. Furthermore, the answers displayed extreme skepticism towards the scientific evidence of the link between meat consumption and global climate change. Rather, those that participated in the study thought that other mitigation efforts involving climate change would be much more effective.

Indeed the consumption of meat contributes to global climate change, and thus spurs discussion of sustainable diets. A sustainable diet consists of food that is nutritionally sufficient, safe, economically fair, affordable, and culturally appropriate with a low environmental impact. The dialogue about such a diet has focused on the environmental and health dimensions, but has largely disregarded the cultural significance and a willingness to change. Eating meat can be a nutritionally a double-edged sword. Meat can be a high source of protein and micronutrients, but has shown to increase risks of some chronic diseases. It is estimated that livestock accounts for nearly 14.5% of the worlds anthropogenic greenhouse gas emissions. Thus reducing one's consumption of meat decreases our contribution to global climate change. In the United Kingdom, 56% of men and 32% of women eat more than the healthy amount of red meat a day, suggesting a serious trend of meat culture in the developed world.

The interviews and focused group discussions were diverse in age, economic status, regional locations, education levels, employment, and family structure. Most of the participants accepted that climate change is a true phenomenon, yet some of these particular participants were clueless about the way food production impacts the environment. Food transportation, packaging, and production were perceived as the central problems more than actual meat production. Once aware of the environmental impacts, a majority of the participants either were skeptical of the information or cynical about their individual impact on reducing climate change by not eating meat. They felt powerless and overwhelmed by the presented information. There did not appear to be obvious differences between sexes, socioeconomic, and regional groups. The opposition to changing their eating habits was reduced to the simple fact of pleasure. The act of eating meat for most was a pleasurable and affordable experience. Thus, why take some form of tradition and happiness away when individuals cannot directly see the changes in their contribution to global climate change? This article presents a unique perspective on the significance of food in our lives and its impact on our futures. In order to combat global climate change, we must invest in innovative technologies and efficiencies in our current institutions. However, we must also seriously look at the sacrifices and changes we must make culturally.

Mindsets about Changing our Beef Consumption

Klöckner (2017) sides with the assumption that food selection is a product of one's socio-economic, political, cultural, and physical conditions as well as cognitive and affective factors like perceptions and preferences. These factors are analyzed and understood by identifying habits and routines that tend to predict actual selection.

The study used the stage modeling approach to understand how humans voluntarily change their behavior in regards to beef consumption. A recruitment letter was issued to a random sample of Norwegians for the first sample. The second sample used an online panel operated by TNS Gallup. The first survey had very little data and was subject to volunteer bias, thus the second sample was surveyed to collect more comprehensive and reliable data. The two surveys were identical, first asking questions about the participant's actual consumption of meat, then stretching into asking participants to choose from a series of statements about mindsets of meat consumption. With statistical analysis, the two tests proved to be significant in their findings.

The initial intention to reduce beef consumption is initiated by personal norms, and then enforced by social norms and awareness of negative consequences. The strongest correlated relationship between goal and behavioral intensions lay within reducing proportion size more than introducing vegetarian diets. The easier the change is to make, the stronger the intention to commit to it. Practicality of accessibility and planning of such substitutions also had a negative impact for implications. When tracking the meat eating tendencies of the participants, beef consumption was noticeably reduced in the final, post action stage.

To be frank, this study gave statistical significance to otherwise fairly common knowledge. People dislike change unless it presents itself as simple, accessible, and not excessively intrusive to their current way of life. People tend to be motivated by self-preservation, whether that literally is the state of their health or status by adhering to social norms and opinion. People tend to lean towards changing their proportion sizes than substituting meat with an alternative. Substituting a vegetarian option for meat was popularly unfavorable. The stage model used in this analysis can be a helpful tool for policy makers, government officials, and business people to influence change. It is unclear if this behavior would be promoted by businesses and livestock agriculture. If anything changing social norms will have to be the backbone to a significant change in

Life Cycle Assessment and Comparison of Grass-Fed Beef and Dairy Production

Moving towards heavy reliance on local food systems may improve the sustainability of beef production. The demand for regional grass-fed beef is high in the Northeast, and these systems have low environmental costs due to their possible multi-functionality. Tichenor *et al.* (2017) analyze and quantify the environmental impacts of management-intensive grazing of grass-fed beef with confinement dairy beef production by using a life cycle assessment. The comparisons of the two identify potential disadvantages and benefits for both production systems in terms of environmental integrity. The bounds of such an assessment included potential of climate change, eutrophication and acidification potential, water depletion, and fossil fuel and land use. The system analysis was cradle to grave, or rather farm to gate. Thus it stretched from feed production and processing, to cattle transport. In general dairy beef production had higher fossil fuel use than grass-fed beef and was more water intensive. However, while grass fed systems resulted in calculated lower impacts for eutrophication and acidification levels, dairy beef production had significantly lower impacts in every other category. Through a sensitivity analysis, Tichenor *et al.* concluded that possible carbon sequestration and lower methane emissions might lower the environmental consequences of grass fed beef. Tichenor *et al.* state that further research and analysis should be aimed towards reducing regional grass fed and dairy beef environmental footprints through holistic approaches, like substituting food waste for conventional feed. Nevertheless, multi functionality of dairy beef to produce both meat and dairy products is statistically shown to be most sustainable.

The US produces nearly 11.8 million kg of beef annually, making it the top producer of beef in the world. Americans consume over 90% of this beef, and cattle emit over half of greenhouse gas emissions by livestock in the US Thus, a life cycle analysis of each system was taken to try to reduce this impact. The assessments were measured regionally due to differing climate, and to emphasize the possible sustainable benefits of local food production and consumption. Tichenor *et al.* clearly defined the characteristics of each system, reached out to farms when collecting data, and focused on feed production, manure, and enteric emissions in addition to the life cycle assessment and sensitivity analysis.

The analysis had several comprehensive and tangible conclusions. Grass-fed and dairy beef had a global warming potential of 33.7 and 12.7 kg respectively of carbon dioxide eq per kg HCW (hot carcass weight). Grass-fed and dairy beef have a total land use of 122 m^2 and 17 m^2 respectively of land per kg HCW. Grass fed beef utilized 1.10 kg oil eq per kg HCW of fossil fuel depletion compared to dairy beef which was 1.33 kg. Eutrophication potential was 0.44 kg N-eq. for grass fed in comparison to 0.18 kg for dairy beef. Grass-fed and dairy beef had acidifying emissions of 30.2 and 12.7 mol H eq. per kg HCW. Maintaining the number of cattle necessary for proper reproduction rates accounted for 60% of system impacts for grass fed, and 52% for dairy beef. General trends illustrated that feed production largely contributed to both grass fed and dairy beef footprints. Thus a substitution to food processing byproducts or food waste into rations opportunistically reduces one's impact. Overall the dairy beef system had lower carbon and land environmental impacts, but higher eutrophication and acidification per land unit than the grass-fed system.

Microbial Proteins are a solution to Sustainable Livestock Feed

Matassa *et al.* (2016) explore the possibility of manufacturing protein-rich feed or additives for livestock. They could be a form of algae, yeast, fungi or bacterial biomass. Such feed would be able to sustain livestock populations with a significantly lower environmental impact. In other words, land and water use of microbial production could be 20 to 140 times lower in environmental impacts than traditional fishmeal and soybean meal. Furthermore, unused agricultural land could be revitalized into a sink for carbon capture.

The global market of livestock feed demands over several hundred million tons of conventional feed per year. Climate change and anthropogenic pressure on the world's natural resources places a major strain on the integrity of agricultural food systems. With an increasing human population, the demand for this feed is projected to simultaneously rise 70% from figures gathered in 2006. Microorganisms, or microbes, are essential in the success of this manufactured feed can be used as direct food sources. Research and development of microbial proteins began in the 1960s. Despite the success and public acceptance for livestock feed, market prices drove consumers to pick the cheaper option of conventional feed like soybeans. Environmental pressures have refocused the market to a resurgence in the development of such feed, and new technology has driven down the price.

Natural gas has been the central method for developing and producing microbial feed. However companies like Unibio, are producing it with innovative fermentation processes producing continuous cultures of *Mthylococcus capsulatus*. The environmental footprint of Unibio's processes is 1000 times smaller than any conventional production system. Aquaculture is the main driver in the development this feed and continues to rely on the production for fishmeal.

Another possible development path for the replacement of soybean production for animal feed is recovering nutrients from the food industry. In this case heterotrophic microbes convert carbon and nutrients waste or processing waters into proteins, as is presently being done in Belgium using potato process water. The industry is attracting new investors with rising environmental threats and changing market conditions. However, the bacterial-based microbe products still relatively lack a significant interest from most buyers.

These microbe products are currently meeting FAO and WHO standards in amino acid storing patterns and thus can also be considered for human consumption. While algae and yeast are the main players, a product called Mycoprotiens is able to reproduce the taste and consistency of meat and is now being sold in 15 countries around the world. Microbial oil can also be used as a suitable replacement for vegetable oil. What stands in the way of this consumption is human perception and acceptance of microbial proteins. However, the potential of this product is fantastic. With increasing environmental fragility, I believe we need to explore avenues like this and simultaneously restructure our food and agricultural systems in order to live more sustainably.

Agricultural Emissions and Climate Change Mitigation Potential

Paustian *et al.* (2016) work to inform political leaders, industry, agricultural producers and the general public about the best form of climate change mitigation. Agricultural practices release carbon dioxide, nitrous oxide, and methane into the atmosphere that add to the greenhouse effect. Thus recent scientific research has been aimed at finding effective climate change mitigation through the agricultural practices and land management. The authors' approach this issue by tackling each gas and using research and data to illustrate more efficient and less emitting agricultural practices.

Using soil as a carbon sink could be extremely beneficial. The storage of carbon within soils depends upon the balance of carbon inputs from plant and animal resides and emissions form decomposition. These inputs and outputs are affected by the climate, physical composition of the soil, and human intervention like agricultural management practices. The absorption of carbon is thus maximized by little soil disturbance and erosion, high water and nutrient efficiency of crop production, and high crop residue return. The most effective carbon sinks have one or more of these qualities. Decreasing the intensity of tilling has been shown to promote carbon sequestration in long-term field studies. No-till compared to highly intensive, conventional tilled annual agricultural systems can increase soil carbon from 0.1 to 0.7 metric tonnes per year. Increased amount of residue returned to soil also aids the rate of carbon sequestration. One can attain such conditions by high-residue yielding crops, hay crop rotations, manure and bio-solids application, and improved management of fertilizer, water and pests.

When pondering the effectiveness of fertilizers, their ability for carbon storage is offset by their production energy inputs, and their potential to increase nitrous oxide and methane. In grazing lands, carbon storage and soil health can be increased by switching from conventional methods to using improved species, sowing legumes, fertilizing, and irrigating. It is essential to restore degrading soils, reforest and afforest, manage desertification, and retire marginal lands. Improving management of agricultural lands should take priority over general fertilization that may increase greenhouse gas emissions. Such management may improve soils and decrease methane emissions from grazing. These practice changes are estimated over two to three decades to capture 80 to 150 million metric tons of carbon or more per year. The long-term affect of management practices will improve soil quality and sustainability for more productive yields. In turn this quality may enhance water storage, infiltration, reduction of runoff and effective storage of plant nutrients, again adding to increased productivity,

Agriculture's contribution of nitrous oxide to human induced climate change stems form nitrification and denitrification that is increased by fertilization. It is estimated that nitrous oxide emissions could reduce by 35% globally by improved crop nitrogen use. Such practices would have greater savings in North America's input-intensive industry. The most impactful mitigation methods for soil emissions associated with nitrogen cycles are fertilizer timing and placement. It is also essential to store and dispose of animal waste in order to reduce the greenhouse gas emissions from grazing animals. A grander solution would be the reduction of the ratio of animals to land.

Ruminant livestock and waste are predominant sources of methane emissions in North America. To reduce emissions from enteric fermentation in livestock, one could add feed additives; specialized bacteria, proportion feed ratios, or overall be more efficient with the production of livestock. With a more efficient system, there is less animal waste but more product per animal. Large scale digesters, lagoon covers, and other waste storage practices can be used to offset and reduce emissions from livestock waste.

Finally, using biofuels in place of fossil fuels in the production process can reduce the dependence and usage of fossil fuels in agricultural production. Dedicated energy from crops, waste and residues from farms can be used in addition to methane from waste. Such products can be used for fuel, power, and chemical feedstocks.

The authors suggest three different policy implementations in order to lower our current emissions from Agricultural production. The first suggestion is international agreements that allow land sinks, including agricultural and forest land, to be recognized and used in climate change mitigation practices. The second is that the US could encourage carbon sequestration in soil, as a measurement for productivity of land. Finally emitters could buy offsetting credits from farmers in order to obtain "climate change- neutral products". The issues with implementing these types of policies are the difficulty in measuring and regulating full greenhouse gas emission rates and sequestration rates and the permanence of each action. Furthermore, sufficient incentives must be provided to aid farmers and the agricultural industry with the large up front costs that come with these practices. However, the increased efficiency and productivity of crops and animals as well as a reduced contribution towards climate change is a notable and significant benefit.

Attributional Life Cycle Assessment and Food Miles

Life cycle assessment (LCA) investigates the environmental impact of a product from cradle to grave. This cycle includes resource extraction, production, transportation, use, and disposal. The assessment covers a wide range of environmental impact categories from smog to global warming. LCA has become widely recognized as valid and thus utilized for policy formation. There has been a shift in preference to ALCA, or accounting-style LCA. ALCA additionally assesses market dynamics and human behavior. However, ALCA has been shown as discountable due to problems with fixed inputs, outputs, and human behavior. Yang *et al.* (2017), demonstrate the possible success of ACLA with modifications in the environmental issues of food miles and changing diets.

Optimal local food consumption describes food systems with few food miles, or distance between production and consumption. There is a current push towards increasing capacity and practice of local food consumption. Drivers include consumer perception of increased quality, better knowledge of food origins, support of local economies, and environmental sustainability. The ALCA of local food consumption illustrate that food miles are a small contributor to climate change with only 11% total greenhouse gas emissions. This finding thus discourages local food production. The authors argue that ALCA measurement uses a national model of transaction and emission data, but does not take into account regional characteristics and aggregates all types of food together. Therefore, ALCA presents a warped perception of the importance of local food consumption.

Local food consumption offers ways to sustainably manage wastewater. Treated wastewater from urban areas in close proximity to farms could be used as irrigation water. The reuse of such water would reduce demand on groundwater, allowing more time for aquifers to replenish. A shared partnership reduces the competition of water between cities and farmland. Furthermore, cropland could be utilized as a means for filtrating basins, which would reduce the intense processes of ensuring water quality in aquifers. In addition, recycling of urban organic waste as compost can reduce the need for fertilizers and improve soil quality. Close proximity of urban and rural lands would also reduce the impact of urban heat islands and their contribution to global warming. The framework of ALCA lessens the perceived environmental impact of human diet change as well. Findings from a German study show that water usage could be reduced by approximately 26% and greenhouse gas emissions could be reduced by 11% if we switched our diets to vegetarian. Yang *et al.* (2017) model diet changes in addition to localization of food systems and find that the US could possibly reduce the blue water footprint by nearly 50%.

The knowledge of the potential impact of these changes should motivate policy makers to incentivize consumers to eat more local produce, and agricultural industries to reduce the range of product transport. Or rather heavily investigate the food miles and benefits of production of particular produce. This study illustrates that ALCA should not be used as the primary motivation for policy making. Thus ALCA does not recognize that although vegetables are more water intensive than meat, we can produce vegetables in less water-intensive ways. The authors specify that this study did not intend to assess the validity and importance of food miles. However, I think that the arguments presented here indicate how drastically our environmental impact could change if we pressured locality in our food systems.

Climate Change Mitigation through a Food Carbon Tax

Agriculture accounts for approximately 24% of total greenhouse gas emissions with the inclusion of land use and fertilizer production. Eighty percent of greenhouse gas emissions stem from the production and consumption of meat and dairy products. García-Muros *et al.* (2017) evaluate Spaniards' current diets, and suggest that a change in consumption patterns is the key to greenhouse gas mitigation. They argue that mass consumption of red meat, sugar, and saturated fats contribute greatly to weight gain, and increase disease risk. To incentivize such a change, they suggest that Spain implement a carbon tax on food. This study goes beyond the success of a carbon tax, and looks at the welfare impacts for different income, age, and social groups in different food categories.

García-Muros *et al.* use an Almost Ideal Demand System model, which makes an approximation about an unknown demand system. The model complies with economic consumption theory and does not limit utility. The model organizes and evaluates data collected from 2002 to 2013 of over 20,000 Spanish household expenditure surveys. The demand model assesses 14 different food categories including beef, chicken, pork, fish, milk, dairy products, eggs, cereals. fruits, vegetables, potatoes and potato based food, oils and fats, sugar and sweet products, and other food products. Price and expenditure elasticities are calculated for each of these categories. Finally three tax scenarios are used. Each assesses welfare and distributional effects of a food carbon tax. The first reference scenario uses a carbon tax based on food with a carbon price of 25 euros per tonne of carbon. The second is a high carbon price scenario with a carbon price of 50 euros. The third is the high carbon price scenario with exemptions for cereals, fruits and vegetables because of their low emissions. Accordingly, animal products have the highest emissions and therefore the highest tax rate.

Depending on the rate of elasticity at the time, these various tax scenarios unevenly increase food prices, modify consumption, and thus reduce emissions. The first tax scenario is projected to decrease emissions by 3.8%, while second would decrease emissions by nearly 7.6%. The final tax scenario would drive down meat consumption significantly, and reduce emissions 4.7% with an increase in vegetable consumption. García-Muros *et al.* use the Hicks model to project welfare effects. These findings show that the third scenario has the possibility of being most cost-effective by having the same carbon price as scenario two, but low welfare impacts as scenario one. The progressivity and redistributive effects model show that low-income houses are consuming greater proportions of the exempted goods in scenario three. Such a change in eating habit demonstrates a reduction in cholesterol, saturated fat, sugar and protein intakes.

Many countries have tried to implement such a tax, including Denmark's "fat tax" and Hungary's "junk food tax". They have been met by critiques emphasizing that distributional impacts and acceptability degrade their effectiveness. Such taxes have greater substantial financial implications on the poor than the rich, and are unclear in their direct impact on public health. Therefore, critiques state that it is questionable if the benefits outweigh the costs. Based on the results of this study, a high carbon tax with exemptions is the most beneficial for human health and the environment with relatively low distributional impacts. Such a tax for Spain and possibly other nations around the globe is essential in reducing the environmental impact of agricultural systems. However, the authors do admit it may be difficult to implement and gain support for a carbon tax on food. I think this study is extremely fascinating, but may not be promoted and implemented by such a strong meat consuming public as the United States. As a country, we need to become more enlightened about the environmental implications of our consumptive patterns.

Willingness to Pay for Food Products with Environmentally Friendly Labeling

Climate scientists recommend an improvement in agricultural practices and production as a solution to reduce greenhouse gases. In addition, a change in agricultural practices is being incentivized by increased pressure from buyers, retailers, processors and government agencies. Akaichi *et al.* (2017) carry out an experiment centered on willingness to pay to explore the validity of this claim.

Rice was the product selected for this experiment for several reasons. First, rice is a main food source for over half of the world's population, and is a source of high greenhouse gas emissions in the agricultural sector. Correspondingly, the countries with the top rice producers are similarly the countries with the highest greenhouse gas emissions. Hybrid rice offers a reduction in greenhouse gas emissions in comparison to conventional rice. In fact, hybrid rice production increases yields between 15–20% with the same inputs as conventional production. Furthermore the appearance of hybrid rice is indistinguishable from its non-hybrid counterpart. This experiment seeks to determine consumers' willingness to pay for hybrid rice compared to conventional rice, and then determines their preference for lower greenhouse gas emissions, locality and food miles.

Akaichi *et al.* used Wells conventional rice and XL723 hybrid rice. It was estimated that Wells conventional rice produced about 9.97 oz. per pound of CO_2 emissions, while XL723 hybrid rice produced about 8.21 oz. per pound. The experiment auctioned off four types of rice in this example sold in Arkansas. The first was conventional Wells produced in Stuttgart, Arkansas with 250 food miles. The second was XL723 produced in Stuttgart with 250 food miles. The third was Wells from New Madrid, Missouri with 422 food miles. The fourth was XL723 also from New Madrid, Missouri with 422 food miles. The experiment used the US Congress's definition of 'local food', or rather food produced and transported within a state's borders. In the auction-based experiment, participants were asked to report their willingness to pay for each type of rice. Participants were provided with financial incentive at the beginning of the experiment, and a detailed oral explanation of how the procedure and auction would work. After all questions were answered, a training session was carried out before the actual experiment. The participants then physically examined the products. Seven treatments were given for a set of 50 participants each. Each of the seven treatments were presented in a different order, but included the categories greenhouse gas emissions, origin, food miles, taste, and no information at all.

With no labeling whatsoever, consumers were willing to pay a price premium of 11% for hybrid rice based on appearances alone. The preference significantly increased when consumers were informed the hybrid rice was environmentally friendly with lower greenhouse gas emissions. Furthermore, without the information of lower greenhouse gas emissions, consumers still increased their willingness to pay significantly for local origin and lower food miles. Thus when the hybrid rice was shown to have higher food miles than conventional rice, their willingness to pay decreased for the hybrid. By using a between-subject analysis, the results demonstrated that the information on reduced greenhouse gas emissions was much more effective in increasing consumers willingness to pay than food miles or origin. Furthermore, consumers are more willing to buy a hybrid rice labeled with lower greenhouse gas emissions than conventional rice with lower food miles. Without the knowledge of greenhouse gas emissions, consumers had a higher willingness to pay for the product with lower food miles and local origin. The results also showed that the concepts of origin and food miles were not interchangeable. As expected, participants of similar socio-demographic characteristics had similar opinions and responses with labeling. Older participants had notably lower price premiums for local rice with lower greenhouse gas emissions and food miles, and higher educated participants were willing to pay a greater price.

I found this experiment emphasized the increase concern for global climate change, and the importance of labeling sustainable products. People are willing to pay more for a good that is environmentally friendly. Thus labeling these products can promote and increase in sustainable farming practices. However, it was interesting to see how greenhouse gas emissions automatically took precedent over locality and lower food miles. I am curious to see this experiment further developed with different food products.

Conclusions

After reviewing the articles, I come away with the conclusion that agriculture needs to be a key topic when discussing climate change. The public should be further educated about the immense environmental consequences of consuming meat, and especially beef. A wider education on agricultural contributions to global warming could spark a greater shift in national and global food culture. It may become less socially acceptable to consume mass amounts of beef and other meat products. Moreover, those who are educated in food and agriculture practices would recognize the contribution of low meat consumption, utilizing locally produced food, and lowering food miles to reduce greenhouse gases in food production. It is essential that policy makers recognize the value of a transformation in our food consumption and production in order to reduce our contribution to global climate change.

References Cited

Akaichi, Faical, Nayga, Rodolfo M., Nalley Lawton Lanier. 2017. Are there trade-offs in valuation with respect to greenhouse gas emissions, origin and food miles attributes? European Review of Agricultural Economics 44, 3–31.

Danish Agriculture and Food Council, 2017. Danish Pig Meat Industry. Web.

García-Muros, Xaquin, Markandya, Anil, Romero-Jordán, Desiderio, González-Eguino, Mikel. 2017. The distributional effects of carbon-based food taxes. Journal of Cleaner Production 140, 996–1006.

Klöckner, Christian A. 2017. A stage model as an analysis framework for studying voluntary change in food choices - The case of beef consumption reduction in Norway. Appetite 108, 434–449.

Macdiarmid, Jennie I., Douglas, Flora, Campbell, Jonina. 2016. Eating like there's no tomorrow: Public awareness of the environmental impact of food and reluctance to eat less meat as part of a sustainable diet. Appetite 96, 487–493.

Matassa, Silvio, Boon, Nico, Pikaar, Ilja, Vertraete, Willy. 2016. Microbial protein: future sustainable food supply route with low environmental footprint. Microbial Biotechnology 9, 568–575.

Paustian, Keith, Babcock, Bruce, Hatfield, Jerry *et al.* 2016. Agricultural Mitigation of Greenhouse Gases: Science and Policy Options. Virginia Tech: College of Agriculture and Life Sciences.

Tichenor, Nicole E., Peters, Christian J., Norris, Gregory A., Thoma, Greg, Griffin, Timothy S. 2017. Life cycle environmental consequences of grass-fed and dairy beef production systems in the Northeastern United States. Journal of Cleaner Production 142, 1619–1628.

WWF, 2014. Living Planet Report 2014 Summary. Encyclopedia of Corporate Social Responsibility, 1-36. Web.

Yang, Yi, Campbell, J. Elliott. 2017. Improving attributional life cycle assessment for decision support: The case of local food in sustainable design. Journal of Cleaner Production 145, 361–366.

Ocean Acidification

Elizabeth Rodarte

Ocean acidification is a process that decreases the pH of seawater throughout the world's oceans caused by the uptake of atmospheric CO_2. Oceans act as a carbon sink by taking in atmospheric carbon dioxide, both natural and anthropogenic. Anthropogenic emissions stem from fossil fuel burning, deforestation, and other processes while natural sources of CO_2 include animal and plant respiration, decomposition of organic matter, forest fires, and emissions from volcanic eruptions. The increase in carbon dioxide into the ocean disturbs seawater chemistry as well as the distribution of nutrients relative to latitude and temperature. As CO_2 enters the ocean it bonds with carbonate $[CO_3^-]$ and water to form bicarbonate ions. This process adds H^+ ions thus lowering the pH and replacing potential calcium to bond with carbonate. Important carbonate nutrients in the oceans are aragonite and calcium carbonate. The process of ocean acidification limits calcification of marine organisms such as crustaceans, mussels, and coral reefs that depend on calcium carbonate for their shells. Aragonite and calcium carbonate saturation is high in the tropics and decreases towards the poles. The temperature in the tropics enhances the dissolution of these nutrients. Higher latitudes have lower concentrations due to the cold water decreasing solubility. Thus, ocean acidification has more dramatic effects at the poles than in the tropics. It is an on going problem that humans cannot see themselves on land, but can easily see the effects when looking at coral reef and wetland ecosystems. The implication of ocean acidification is that it not only affects marine organisms and ecosystems, but humans as well. Climate change since the industrial revolution has weakened the chemistry of the oceans by unnaturally increasing the amount of CO_2 uptake as well as disturbing the natural balance of nutrients with feedback loops of increased temperature. Many studies have been conducted to investigate the specific effects of ocean acidification on marine organisms and habitats. Other studies focus on how humans are a driving force to these diminishing resources, as well as how this could have occurred in the distant past. With more information, scientists can better identify which actions have the most influence to ocean chemistry and possibly find a way to mediate, or even reverse the effects of ocean acidification.

Ocean Acidification Can Mediate Biodiversity Shifts by Changing Biogenic Habitat

Ocean acidification is the process in which the pH of the world's oceans decreases due to the production of atmospheric CO_2. The increase of CO_2 and decrease in pH leads to changes in calcification, growth, and abundance of species such as coral reefs, mussels, seagrass, and macroalgae. Habitats experience the indirect effects of such CO_2 increases. They must remain resistant to sudden changes in pH and CO_2 in order to benefit the organisms they support. By modeling the effects of lowering pH in habitats with corals, mussels, seagrass, and microalgae, we can determine the costs to these species. Coral reefs and mussels are calcifying organisms that are negatively affected by lowered pH, which limits survival and stunts, or even stops, growth and development. Lower pH decreases the species complexity of corals and mussels and ultimately the species richness in habitats. *Mytilus* mussels, for example, require specific pH to function. The species of mussels, other than *Mytilus,* that survive decreases in pH lack "structural complexity" to support dense surrounding vegetation. Therefore, the loss of *Mytilus* mussels due to ocean acidification allows for a more stable yet less diverse habitat (Sunday 2016).

Seagrass and macroalgae differ from coral reefs and mussels in that they benefit from ocean increases in CO_2. These photosynthetic organisms increase their biomass and species richness along with decreases in pH. However, in terms of habitats, the increase in CO_2 leads to a response of "competitive replacement" between species for spatial area. Thus, habitats also experience decreased species richness as species compete with each other for resources. This process takes longer to see than in coral reefs and mussel beds.

Different marine habitats experience a variety of consequences to ocean acidification. The positive side of the consequences of ocean acidification in different habitats allows for the study of "habitat recovery after disturbance." Knowing that ocean acidification affects each habitat and species differently can better prepare us to treat habitats more efficiently.

Macroclimatic Change Expected to Transform Coastal Wetland Ecosystems this Century

Wetland ecosystems are exceptionally valuable, both ecologically and economically. They provide services such as storm protection, nutrient removal, carbon sequestration, erosion prevention, and wildlife habitat. Wetlands have been diminishing around the globe without an agreed reason. Gabler *et al.* (2017) examine the effects of macroclimate on wetland ecosystems. Macroclimate (temperature and precipitation) determines the type of wetland conditions in a certain area whether it be warm and wet mangrove forests, cool and wet graminoid marshes, arid succulent marshes, or extremely arid and saline unvegetated flats. The four types of wetlands are structured by macroclimate; thus it is hypothesized that severe changes in macroclimate have direct influences on wetland vegetation, biomass coverage, salinity, and mangrove presence. Grabler *et al.* gathered extensive field data from 1981–2010 of 10 estuaries at various temperatures and precipitation rates in the Northern Gulf of Mexico. They created two models to predict the dominance of functional groups and vegetation structures, finding that there is a strong link between vegetation and precipitation. The graphed models indicate sigmoidial relations to minimum surface air temperature and mean annual precipitation. Graminoid marshes and mangrove forests have thresholds as to how much they can resist changes in surface air temperature until one functional group outcompetes the other. The same can be said about unvegetated flats and succulent marshes. The ten estuaries studied presented a gradually positive relation between precipitation and vegetation height while there was a predominant shift for minimum surface air temperature.

Models were made to show changes in wetland ecosystems in response to increasing temperature. The effects of increasing temperature include increased erosion, negative responses to sea level rise, and decreased vegetation height. These models can be useful to "inform conservation and restoration efforts" to regions beyond the Northern Gulf of Mexico.

Humans as a Hyperkeystone Species

Keystone species are organisms that have larger ecological impacts than others of their biomass. Worm & Paine (2016) identify humans as "hyperkeystone" species that affect both terrestrial and marine ecosystems, and examine the human effects to organisms around the globe. The complex interactions are best displayed through food web charts that describe the change in habitat, behavior, and pollution in an ecosystem. From the data, food web theory explains why carnivores in aquatic food webs and herbivores in terrestrial food webs have the most keystone species. Worm and Paine also address the "shifting baseline problem"; as a large carnivore becomes rarer in terrestrial ecosystems through fishing, hunting, or nontrophic impacts, its ecological role becomes larger. Exploitation rates humans exert on herbivores, carnivores, and top predators were graphed based on types of aquatic and terrestrial ecosystems. Aquatic ecosystems displayed greater exploitation rates for all trophic groups. The impacts humans have on keystone species can cause trophic cascades such as with sea urchins and kelp, starfish predation, and mussel and algae populations, as well as the interaction between salmon, nutrients, and riparian forests. Of course the human hyperkeystone impact depends on which other species are also considered keystone species within an ecosystem. The role of humans as hyperkeystone species has not changed. It dates back to the Pleistocene, in which occurred an 'overkill' of vegetation. What has changed over time are the vectors of the effects and the measurable consequences. The effects humans have on ecosystems stem from climate change, ocean acidification, and change in keystone species. Anthropogenic noise is one form of pollution that affects many species, both aquatic and terrestrial. Focusing on food web mechanisms and bridging the gap of measuring humans and ecosystems as externalities helps solve the questions of why some communities are more negatively impacted than others. By doing so, trophic cascades can be avoided in the current epoch of the Anthropocene.

Acidification of East Siberian Arctic Shelf waters through Addition of Freshwater and Terrestrial Carbon

The Arctic seas are exceedingly susceptible to ocean acidification due to CO_2 dissolving more easily than in warmer tropical waters. However, atmospheric CO_2 is not the only cause of rapid acidification in the Arctic. The Eastern Siberian Arctic Shelf (ESAS) is estimated to become twice as acidic by the end of the century due to river discharge, ice melt, and terrestrial organic matter. Data from the Eastern Siberian Arctic Shelf in the years between 1999 and 2011 were collected measuring aragonite saturation, inorganic carbon, salinity, and $\delta^{18}O$ to suggest that ocean acidification is driven by terrestrial organic carbon and discharge of Arctic river water and not necessarily atmospheric CO_2. Aragonite saturations, Ω_{ar}, are measured in terms of calcium and carbonate relative to concentrations of solubility at certain temperatures, pressures, and salinities. Ω_{ar} greater than one are useful for calcifying organisms that rely on $CaCO_3$ for their shell and skeleton development. Ω_{ar} less than one dissolves organisms' skeletons by depleting the amount of available CO_3 in the ocean. The ESAS has 4 river sources that meet and erode the coastline by depositing terrestrial inorganic carbon. Ω_{ar} is measured by collecting the seawater Ω_{ar} for aragonite. Terrestrial organic matter drains in to the ocean by eroding organic material in the permafrost. It was found that there was a strong relationship between decreasing Ω_{ar} and increasing pCO_2 and lower pH. A weak correlation occurs between, pCO_2 and O_2, as well as pCO_2 and primary productivity.

Sea-ice Transport Driving Southern Ocean Salinity and its Recent Trends

Salinity plays a key role in ocean water characteristics. Saltier water is denser than fresh water and thus sinks below it, creating layers. These layers can be divided up into deep water, bottom water, intermediate water, and surface water with denser saltier water on the bottom. Movement of Antarctic sea ice is a major contributor to the changing salinity of the Southern Ocean. As global warming intensifies, glacial sea ice is melting at a concerning rate. During freezing, salt is not included in the ice, making it fresh and leaving the sea water saltier. Wind driven fresh water is transported from the Southern Ocean by about 20% between the years 1982–2008 (Haumann *et al.* 2016). The Southern Ocean of Antarctica displays noticeable effects of climate change in the oceans with the increase of fresh water melting freshening the water changing its salinity and pH. Antarctic Bottom Water (AABW) and Antarctic Intermediate Water (AAIW) are mixing less well with the denser Circumpolar Deep Water (CDW) due to the intensified stratification of the water column. This stratification also makes less likely for the CDW to rise the surface and upwell nutrients. The reduce in upwelling has negative effects for marine life similar to the negative effects of the El Niño event in the Pacific. Fish populations die and nutrients are prevented from traveling to other oceans. Unlike El Niño, the melting of sea ice is constant and has been increasing for decades (Haumann *et al.* 2016). Anomalies have been measured and mapped to display an upward trend in Northward sea ice freshwater transport since 1982.

The Southern Ocean plays a large role as a climate change regulator and carbon sink. With the increased stratification, it will intake more atmospheric CO_2. This intake can be harmful to the ecosystems as it contributes to ocean acidification by reducing the pH and making carbonate and aragonite insoluble for calcification. Haumann *et al.* calculated a freshwater flux rate of -0.02 ± 0.01 g kg⁻

[1] per decade. This is detrimental to the global ocean circulation system as well as marine ecosystems due to the magnitude of increased salinity that follows.

Quantifying the Volcanic Emissions which triggered Oceanic Anoxic Event 1a and their effect on Ocean Acidification

An oceanic anoxic event occurs when deep ocean water is depleted of O_2 to the point of not being able to support life. Ocean-wide anoxic events can lead to acidification of ocean water, decrease in global oceanic diversity, and decrease in overall productivity. Bauer, Zeebe, and Wortmann (2017) conclude that volcanoes act as triggers for these ocean wide anoxic events. Bauer *et al.* create two models to determine the magnitude of the volcanoes that erupted in the cretaceous period along the Ontong Java Plateau. The first model is the Os box-model which calculates the magnitude of volcanism to produce the amount of $^{187/188}Os$ recorded in the sediment layers of black shale. The second is the C-cycle model, know as the Long-term Ocean-atmosphere Sediment Carbon cycle Reservoir Model (LOSCAR). The C-cycle model also calculates the magnitude of volcanism needed to produce the amount of $\delta^{13}C$ as well as the carbonate compensated depth (CCD) and aragonite/calcite saturation (Ω_{ar}). The Os and C-cycle models determine volcanic activity necessary to produce the $\delta^{13}C$ and $^{187/188}Os$ that coincide with ancient anoxic events. Both models are used and compared in the study to better the precision of the measurement of volcanic magnitude and CO_2 emissions. The carbon isotopes serve as a proxy for the amount of organic carbon $\delta^{12}C$ and atmospheric carbon pCO_2. From the box models, a negative carbon isotope excursion emerged thus leading to non-linear feedback loop. Volcanic activity increases atmospheric pCO_2 which stimulates weathering of PO_4. The increase in nutrients to the ocean stimulates surface productivity enough to trigger anoxic deep water and the emission of more PO_4.

The data from the two models suggest that the volcanism triggering the Oceanic Anoxic Event 1a was "rapid and intense". The volcanism had a six-fold increase in strength over a 50 kyr period causing severe increases in pCO_2 and CCD as well as severe decreases in $\delta^{13}C$ and Ω_{ar}. The ocean acidification decreased the amount of calcification and calcifying organisms over the 50 kyr period of Oceanic Anoxic Event 1a. This study indicates that global climate change is not the only trigger for global events such as ocean anoxic events and acidification. Volcanism has a greater impact on ocean productivity than one might have previously thought.

The Integral Role of Iron in Ocean Biogeochemistry

Iron is a micronutrient important in the global ocean biogeochemical cycles and primary productivity. Although insoluble, iron has unique ties to the oceans' carbon and nitrogen cycles. Tagliabue *et al.* (2017), show that increases in iron have lead many to view the ocean iron cycle as important to organisms such as phytoplankton. Iron circulates from eroded dust from continents and from hydrothermal activity. It is then upwelled with other nutrients that together stimulate biochemical activity and circulates far beyond its source. The uptake of iron increases primary productivity and a decrease in atmospheric CO_2. A balanced system is beneficial to ocean chemistry and ecosystems, however, increased iron and CO_2 can lead to negative feedback loops of increased productivity causing algal blooms and ocean acidification from the CO_2. The productivity of high and low latitude regions is also moderated by the availability of nitrates that are required for permits the growth of non-nitrogen-fixing organisms. Organic iron-complexing ligands do not vary much in abundance, but they do however, control the ocean interior iron regeneration. These ligands have a larger influence on contemporary atmospheric CO_2 levels and influence the distribution of dissolved iron.

The coupling effects of iron relative to other nutrients can be evaluated from phytoplankton who use both as resources. Southern Ocean phytoplankton, for example, exhibit lower levels of iron than those of the Atlantic Ocean. Thus, one can conclude that there is more iron in the Atlantic as opposed to the Southern Ocean. Models are currently being used to determine future effects of iron cycling relative to climate change and changing ocean circulation, seasonally and in the distant future. More can be learned about the specific biological, chemical, and ecological effects the iron cycle has on the world's oceans. With more information we can better predict consequences to larger carbon and nitrogen cycles from natural and anthropogenic events.

Reversal of Ocean Acidification Enhances Net Coral Reef Calcification

About one fourth of the atmospheric CO_2 emitted is taken in by the oceans each year. Though humans cannot directly see the effects on land, the surplus CO_2 causes and decreases pH, carbonate saturation [CO_3^{2-}], and aragonite saturation (Ω_{arg}), all of which are necessary for community calcification in coral reefs. Coral reef ecosystems are the most vulnerable to ocean acidification effects because the coral organisms heavily rely on the supply of minerals of carbonate and aragonite that become unsaturated at low pH. Because temperature, pH, and nutrient saturation has changed dramatically since preindustrial times, Albright et al. (2016) compared the net calcification response to alkalinity enrichment of preindustrial conditions. They tested One Tree Reef, a pseudo atoll in the Great Barrier Reef, and used its three lagoons to separate the flow of water from upstream of lagoon one to downstream of lagoon three. On 15 of 22 experimental days, a tank in the first lagoon released NaOH that traveled to the test site of a mixed reef community with 17% live reef. The increase in alkalinity increased the pH of the test site to simulate preindustrial alkalinity conditions of seawater. A dual tracer method was used by releasing an active and passive tracer. The active tracer was the alkalinity solution while the passive tracer was a nonreactive rhodamine dye. The ratio of the two was used to compare alkalinity uptake of a control and treatment, thus test samples were taken both upstream and downstream, then analyzed for alkalinity, rhodamine, pH, dissolved inorganic carbon, and nutrients. Albright et al. concluded that there was a positive correlation between alkalinity and nonreactive dye both upstream and downstream. The amount of Ω_{arg} increased an average of 0.6 units. The upstream and downstream slopes were significantly different on days of alkalinity and control days. Thus, the null hypothesis that reef calcifiers did not respond to alkalization was rejected. The alkalization was a significant influence more than other multiple factors such as light and temperature in this study that also effect nutrients and net community calcification. The authors were able to state with authority that there is a significant increase in net community calcification of this coral reef when sea water chemistry was set to preindustrial era conditions. As a result of this study, deliberate alkalization has been proposed to offset the effects of ocean acidification on marine ecosystems. Such treatments could be implemented in highly localized areas that have experienced high rates of ocean acidification.

Conclusions

The world's oceans are dynamic and efficient at adjusting to natural changes in nutrient and chemical composition. However, anthropogenic influences have tested the resilience of oceanic systems and marine ecosystems. Increased atmospheric CO_2 has detrimental effects to the calcifying organisms and entire ecosystems. Lowering pH and greater temperatures are pushing the oceans to a breaking point we can already see. Scientific research continues to investigate the anthropogenic stressors of calcifying organisms and the long-term effects of ocean acidification on marine ecosystems.

References Cited

Sunday, J.M., Fabricius, K.E., Kroeker, K.J., Anderson, K.M., Brown, N.E., Barry, J.P., *et al.*, 2016. Ocean Acidification can Mediate Biodiversity Shifts by Changing Biogenic Habitat. Nature Climate Change 7, 81–85

Gabler, C.A., Osland, M.J., Grace, J.B., Stagg, C.L., Day, R.H., Hartley, S.B., Enwright, N.M., From, A.S., Mccoy, M.L., & Mcleod. J.L., 2017. Macroclimatic change expected to transform coastal wetland ecosystems this century. Nature Climate Change 7, 142–147.

Worm, B., and Paine, R.T., 2016. Humans as a Hyperkeystone Species. Trends in Ecology & Evolution 31, 600–607.

Semiletov, I., Pipko, I., Gustafsson, Ö., Anderson, L.G., Sergienko, V., Pugach, S., Dudarev, O., Charkin, A., Gukov, A., Bröder, L., Andersson, A., Spivak, E., and Shakhova, N., 2016. Acidification of East Siberian Arctic Shelf waters through addition of freshwater and terrestrial carbon. Nature Geoscience 9, 361–365.

Haumann, F. A., Gruber, N., Münnich, M., Frenger, I., and Kern, S., 2016. Sea-ice transport driving Southern Ocean salinity and its recent trends. Nature 537, 89–92.

Bauer, K.W., Zeebe, R.E., and Wortmann, U.G., 2017. Quantifying the volcanic emissions which triggered Oceanic Anoxic Event 1a and their effect on ocean acidification. Sedimentology 64, 204–214.

Tagliabue, A., Bowie, A. R., Boyd, P. W., Buck, K. N., Johnson, K. S., and Saito, M. A., 2017. The integral role of iron in ocean biogeochemistry. Nature 543, 51–59.

Albright, R., Caldeira, L., Hosfelt, J., Kwiatkowski, L., Maclaren, J. K., Mason, B. M., Nebuchina, Y., Ninokawa, A., Pongratz, J., Ricke, K. L., Rivlin, T., Schneider, K., Sesboüé, M., Shamberger, K., Silverman, J., Wolfe, K., Zhu, K., and Caldeira, K., 2016. Reversal of ocean acidification enhances net coral reef calcification. Nature 531, 362–65.

Improving Cities to Fight Increasing Urban Temperatures

Deniz Korman

As global keep temperatures keep increasing at a steady pace, it is likely that certain parts of the globe will be facing serious temperature-related human health impacts. Our urban cities are going to be greatly affected by this upcoming issue, not only because they are home to a majority of the human population, but also because they experience elevated temperatures compared to their rural counterparts. If we do not take action, this problem can lead to previously unseen mass immigration events from high temperature regions. However, there are ways for us to improve the infrastructures of our cities in order to make them cooler and more habitable, such as expanding urban green spaces.

In this article, I look into recent studies that investigate how we can effectively improve the green infrastructures of our cities to make our cities as habitable as possible in the face of increasing global temperatures. Although we currently have green spaces in most of our cities, we have yet to start integrating them for cooling reasons, and their cooling potential remains an area of active research. Some of the subjects that I present on are cooling efficiencies of different tree species (Lanza *et al.* 2016), how the size and shape of a park can influence its cooling effect (Jaganmohan *et al.* 2016), how we can use these green spaces to also promote biodiversity conservation (Fischer *et al.* 2016), and alternate strategies such as using roof spaces for cooling purposes.

Although these studies attempt to find general trends that will be applicable to all urban cities, most studies conducted within this area of research are limited to narrow geographic regions. However, there exists a practice that has promised to minimize health-concerns via cooling, regardless of the geographic location of the city, and that is targeting vulnerable populations (Vargo *et al.* 2016). While most studies in this field focus simply on temperature-based cooling effects, if our goal is to make our cities more habitable, then we should also keep in mind the need to maximize the health impact of our changes when we shape our cities for the future.

Designing Urban Green Spaces for Optimal Cooling Effects

Climate change is becoming more noticeable every year, and urban areas specifically tend to reach higher temperatures due to their emissions and lack of forestry. This phenomenon is called the 'Urban Heat Island' (UHI) effect, and can result in urban areas that are 4°C warmer than their rural counterparts. As global temperatures keep increasing, temperatures in cities might reach dangerous levels, resulting in health problems and drops in work productivity. An effective way to counteract the UHI effect is increasing the area of green spaces in our cities, which comes with the added advantage of providing recreational spaces and cleaning air. However, we still do not know how to effectively integrate green spaces into urban environments to provide a significant cooling effect.

That is why Jaganmohan *et al.* (2016) have investigated various parks and forests within the city of Leipzig, Germany, to get a better understanding of how they influence the temperature of their surroundings. The city of Leipzig is home to many forests and parks, as well as residential districts, making it an ideal candidate for understanding how green space characteristics (such as size, shape, distance from city center, percentage of tree/shrub cover, etc.) influence their cooling effects. In order to measure the temperature impact of green spaces, the authors randomly selected green spaces, identified their characteristics using aerial photography, and took air temperatures along a transect, starting from the boundary of the green space to 500 meters into an adjacent residential street. They were then able to create different measures of cooling effects, and identify which characteristics correlated with the strongest cooling effects. Their results indicate that forests provide a higher cooling effect than parks, and they found that the influence of size is not simply linear as they expected; the increase in the cooling effect as area increases is stronger for parks than forests, and the shape of green spaces can have an impact on their cooling effect as well. They also investigated the impact of percentage tree/shrub coverage and sizes of bodies of water located within the green spaces, and found that both of them provided significant cooling. Additionally, they also looked into qualities of the areas surrounding green spaces, and found that the presence of trees or shrub coverage in the surrounding streets increased cooling effect of the green spaces, while increasing building density had a negative effect. Surprisingly, they also found that not all urban forests and parks had a cooling effect on their surroundings. This further stresses the importance of understanding the cooling effect of green spaces, as it shows that correct design is crucial for introducing green spaces with optimal cooling effects.

Improving Blue and Green Infrastructure to Counteract Increasing Urban Temperatures

Urban areas experience higher temperatures compared to rural areas, and it is likely that this will lead to health risks within urban communities due extreme heat in the future. However, we have the power to minimize this effect by improving the infrastructures of our cities. An effective way to lower urban temperatures is increasing vegetation and water surface areas, which also provide the added benefit of increasing urban biodiversity, and improving air quality. While this known to be a valid strategy, the magnitude of the climate impact that such an improvement will have when applied on a city scale is unknown. Žuvela-Aloise *et al.* (2016) have modeled the potential of improving green and blue infrastructure within Vienna, and identified the ways in which changes should be applied in order to counteract urban warming as effectively as possible.

They have constructed two models: A simple model that used temperature data to find a trend in overall temperature increases at various locations across the city, and a detailed model that took detailed climate and land-use data to get an understanding of how elevation, density of urban development, as well as air flow affected temperatures across the city. The authors have then used their models to simulate the impact that introducing or expanding green and blue surfaces will have on local temperatures. Their findings suggest that the most effective way to reduce heat load is to introduce green and blue spaces within the areas with highest building densities. Considering that this may be unfeasible, they have also suggested the introduction of green spaces on a wider scale, such as along highways, to be an effective cooling method. Overall, they have found that making small changes such as increasing green and blue spaces by 20%, and decreasing building and pavement density by 10% and 20% respectively would provide a substantial cooling effect, reducing the number of summer days (defined as days in which the maximum temperature is above 25°C) by at least 10 days.

Identifying Suitable Trees for Urban Heat Management

Global ambient temperatures keep rising year by year, and urban areas experience higher temperatures compared to rural areas due to lower vegetation coverage and increased emissions. An effective strategy to counteract this problem is to expand green spaces and improve urban forestry. However, it is important to ensure that the greenery that we integrate into our cities can withstand changing climate conditions as ambient temperatures keep increasing at a rate faster than ever. Lanza and Stone (2016) focus on how global warming has affected the climate conditions around 20 highly populated metropolitan areas in USA, and the impact that this has had on present tree species. They have identified the metropolitan cities that have experienced the greatest rate of temperature increase in the past 10 years, and looked into how the climate conditions around these metropolitan areas have changed between 1961 and 2010. Using these data, they have tested whether climate changes have made the regions unlivable for previously present tree species, and created a future projection that looks into how much species loss would occur if the historical climate shifts re-occurred. They found that within the past 50 years, 6 of the 20 metropolitan areas studied lost one or more tree species. All six of these cities were located in the southeast region of US, which they have attributed to disproportionate growth of metropolitan areas within this region. However, their future projection predicts that 5 more metropolitan areas be added to this list if climate changes follow a similar trend, and these additional cities are spread across the country, rather than being limited to a region. These findings suggest that southeastern metropolitan cities should be our priority for improving heat management, as they are the ones that are experiencing the most biological loss due to climate changes. In addition to this, the climate projections and tree species presented in the study act as a valuable reference in order to ensure that our new urban green spaces survive ambient temperatures that are expected to increase in the following decades. While the scope of the paper was limited to tree species, the climate projections can be used when planning different kinds of green spaces such as vertical or rooftop gardens or urban agriculture.

While introducing resistant trees is a smart idea, we should keep in mind that we need to balance between introducing non-native species and trying to preserve the extant ecosystems as climate conditions keep drifting away from their historical states. Overall, this study touches on a problem that has not yet been explored or taken into account within urban greenery planning, and is bound to become more important as global temperatures keep increasing exponentially.

Evaluating the Impact of Vegetation and Cool Roofs on the Urban Heat Island Effect in Singapore

It is well established that urban areas experience higher temperature conditions than their non-urban counterparts due to a higher density of buildings. However, although the UHI has been well investigated in mid-latitudes, much of the predicted urbanization and urban population is expected to occur in tropical regions of the globe. In addition to this, since tropical climates experience much lower seasonal temperature variation, problems related to increasing temperatures will be felt year round, posing a serious threat. However, it is possible to combat this by improving our urban infrastructures to mitigate urban-specific increased temperatures.

In their 2016 study, Xian-Xiang and Norford used a climate model-based approach to assess how increased vegetation and cool-roof coverage would decrease urban temperatures in Singapore. Singapore is one of the cities leading in employing green-roofs with over 60 ha of coverage throughout more than 500 buildings. It is also in the process of experimenting with cool-roof technology; incorporating surfaces with high solar reflectance that greatly reduce heat absorption by buildings or pavements. This study, which was directed towards measuring the effectiveness of these mitigation strategies, found that increasing green vegetation had a more meaningful effect than increasing reflectance. Their increased vegetation modeling scenario showed temperature reductions up to 2°C in the industrial regions of the city, where the nocturnal UHI warming can reach 2.4°C. On the other hand, their cool-roof scenario showed a cooling effect that was limited to daytime. Since the study was based on modeling, the authors note that their assumptions about how these two strategies might get integrated in real life might be a little optimistic. However, they also noted that the integration of these mitigation strategies might have a greater cooling effect by additionally lowering air conditioning usage. Overall, they suggested that in the case of Singapore, the green vegetation did a better job of mitigating urban heating than did cool roofs.

How Urban Settlement Types and Sizes Influence Plant Biodiversity

Understanding how our cities influence the surrounding plant biodiversity plays a crucial role in developing sustainable and eco-friendly urban planning practices. Plant dispersal and growth occur much differently in urban regions than in rural areas due to the highly heterogeneous availability of habitats and human disturbances. In addition to this, the Urban Heat Island effect can lead to higher temperatures and impact flora. Although previous studies looked at how species richness varies across different sized urban and rural areas, studies that examine how these effects influence the species composition have been lacking.

In their 2017 study, Ceplová *et al.* studied the influence of the size of cities, towns, and villages on the number and composition (origin and residence time) of plant species present. Additionally, they also tested whether there is a meaningful difference between the floras of the different settlement types due to the UHI effect. In order to answer these questions they collected plant diversity data from 1-hectare plots spread across settlements of different sizes within Germany, Austria, Slovakia, and the Czech republic.

They found that overall, species richness scaled with the size of the settlement, as larger settlements had a higher habitat variety. When they looked at the species composition, they found that they were mostly natives and neophytes (non-native species that were introduced after 1500 AD.) Although, they expected the neophytes to dominate the native flora through dispersal by human activities and competition, they found that native plants formed a large portion of the urban flora. They also confirmed the pattern that species richness increases from city squares and boulevards to less urbanized habitats found in the residential areas and peripheries of the urban region, which has been observed for other wildlife such as birds and butterflies. This suggests that species composition in urban habitats is primarily dependent on habitat type. They found something similar regarding the impact of the Urban Heat Island effect, as they did not observe differences in species compositions directly due to the UHI effect, but rather they found that habitat type was the main factor that influenced the presence of thermophilous species. In addition to this, they observed that city centers had a low level of species turnover, which means that they have more homogeneous and low-diversity floras. Although they did not address this as being problematic, the lack of diversity makes these regions more vulnerable to environmental changes such as increasing global temperatures. Overall, their study revealed how different types of plant species are influenced by the size and type of urban settlements, giving us insight into how we can construct more sustainable and eco-friendly cities.

Biodiversity Conservation via Green Spaces in South American Megacities—Much Different Than Their European Counterparts

City growth can have a drastic influence on the surrounding wildlife. However, we have the power to ensure the conservation of local species by integrating them into our cities via parks and forests. In doing so, we need to be careful with preserving biodiversity, as biodiversity plays a key role in ensuring the stability of ecosystems. Numerous studies have been done on biodiversity within urban environments in temperate regions, specifically Europe, but Latin American cities have been generally understudied.

That is why Fischer *et al.* (2016) have decided to look into how biodiversity is shaped in parks of Santiago de Chile. Santiago is a rapidly growing city located in a global biodiversity hotspot that also maintains a high density of introduced species making it an ideal study case. In their study, the authors investigated plant biodiversity, percentage of introduced species, habitat types (grassland and wooded areas) of various parks across the city, and compared their results to studies from other continents.

They found that introduced species made up a majority in the parks, with a mean proportion of 94.5%. They found these numbers to be significantly higher than their European counterparts, and found that most of these introduced species originated mostly from Europe, and 79% were shared with European urban grasslands. Similarly, unlike findings from European cities, the location of a park in relation to the urban-rural gradient did not significantly predict its biodiversity. Rather the authors found that human population growth in the park surroundings was negatively correlated to park biodiversity, and did a good job of explaining biodiversity patterns along with park age and area. Although these measures can be used to predict biodiversity, population growth is associated with dynamic urban environments, and factors such as amount of construction, which were not tested for in this study. Future studies should be conducted to better understand how and why human population growth influences park biodiversity in order to successfully use them for conservation. Overall, the authors found that previous results from Europe cannot be generalized to South American megacities, particularly because these parks were implemented using European designs and seed mixtures. This implies that unique and locally appropriate conservation measures such as re-introduction of native species should be applied within South American urban parks in order to foster biodiversity.

Linking Urban Surface Temperatures to Neighborhood Conditions

Urban regions often experience higher temperatures than their rural counterparts due to increased building coverage. However, temperature patterns can also form within a city based on heterogeneous building and vegetation coverage, causing temperature variations greater than 16°C. This means that different neighborhoods can experience different amounts of heat stress and health consequences based on their characteristics. It is common to look at land characteristics to predict surface temperatures, however it is also possible to look at the social characteristics of a neighborhood to understand its surface temperature—poorer social conditions have been found to be related to higher surface temperatures. The land characteristics of a neighborhood are often influenced by its social characteristics, such as income, race and education, and can be used predictively in a similar manner. However, the relationship between land and social characteristics, and how they affect surface temperatures are not well known.

That is why Huang and Cadenasso have investigated the dynamics between neighborhood social conditions and land cover in their 2016 paper. They collected data about surface temperatures, land coverage, and neighborhood demographics such as income and resident race within Gwynns Falls, Baltimore for their study. Using this information, the authors tried to determine whether it is social conditions that drive land cover patterns, or if it is the other way around. In addition, they looked into whether social conditions can influence land temperatures regardless of their relation to land cover patterns. Their results indicate that neighborhood land cover impacts the surface temperatures both directly and indirectly through influencing the social conditions of a neighborhood. This finding indicates that although increasing green areas within a neighborhood will lower its surface temperature, it will also result in an increase in housing costs, resulting in displacement of the current residents.

This means that greening efforts alone will not be useful in making sure that poorer neighborhoods do not get exposed to dangerously high surface temperatures, and that we must be cognizant of the social composition of neighborhoods, as well as the economic impact of introducing green spaces when planning greening projects. This is a very real issue, and might become problematic in the future because the residents of lower income neighborhoods work in sectors that are directly impacted by increasing temperatures, resulting in a loss of productivity. In addition to this, continual air conditioner usage and sophisticated lawn irrigation practices are also not as accessible for lower income neighborhoods, making it harder to cope with increased temperatures. While most studies that attempt to mitigate increasing urban temperatures focus just on land characteristics, this study shows that looking at social characteristics is equally important, and should not be left out when making planning choices.

Identifying and Protecting Vulnerable Social Groups In Face of Increasing Urban Temperatures

Urban areas experience greater temperatures than their rural counterparts due to their high density of buildings. In addition, dense low-income urban populations are exceptionally vulnerable to temperature-related human health impacts. It is possible to counteract increased urban temperatures by integrating more green spaces and increasing the reflectivity of built surfaces. However, if we want to make sure that we minimize temperature-related human health risks using these strategies, we also need to look at the social and spatial distribution of populations to identify the most vulnerable regions.

To address this issue, Vargo *et al.* (2016) have tested the effectiveness of urban cooling strategies to reduce heat-related mortality rates for different income levels, races, and ages in the metropolitan cities of Phoenix, Atlanta, and Philadelphia. They used existing projection models to predict the spatial and social distribution of people, as well as the climate conditions in year 2050. Then, they looked at three different cooling scenarios; one which increased the albedo of pavings and roofings based on current commercial roof cooling and paving products, a second which improved vegetation coverage by assuming certain green coverage levels were met, along with converting roof areas to grass in some regions, and third, a combination of the albedo and vegetation changes. By looking at how much mortality loss each scenario induced for each social group at 2050, the authors were able to identify the regions that would most benefit from the implementation of these cooling strategies. Most of the findings varied across the three cities due to variations in the way that social groups are split up, however, they found that in general, areas with older populations benefited much more from these improvements, so targeting older-neighborhoods for cooling projects could result in meaningful changes in mortality rates. In addition, they also found that both types of cooling strategies proved to have a positive health impact on dense lower-income populations that lived in close proximity of the urban core.

For this study, the authors only looked into temperature projections, and possible improvements provided by cooling projects. However, there are other factors which influence temperature-related health problems, such as housing quality and lack of air conditioning, which can increase the risks of extreme heat. This suggests that low-income neighborhoods are even more vulnerable to climate changes than predicted by their mortality models, and should be prioritized for cooling projects. Even though this study was not able to find strong general trends, it has shown that social factors such as age and income level can, and should be used when implementing urban cooling projects in the future in order to maximize their health impacts.

Conclusions

Assuming that current increasing temperature trends stay constant, we will have to improve our cities in order to keep them habitable. All of the studies discussed above provide us meaningful insight into how we can effectively improve the infrastructures of our cities for the future. However, when making these changes we should be sure to not only focus on the absolute cooling, but other factors such as how we can minimize human-health concerns, and conserve biodiversity. In addition to this, these studies, which were conducted at various locations across the globe, show that cities respond to climate change very differently, and demonstrate that some cities are at greater risk than others. Since making infrastructural changes can take a long time, if we are to be ready for the future, than I would suggest identifying the cities that are the most vulnerable to temperature-changes and planning changes immediately.

References Cited

Čeplová, N., Kalusová, V., Lososová, Z., 2017. Effects of settlement size, urban heat island and habitat type on urban plant biodiversity. Landscape and Urban Planning 159, 15–22. doi:10.1016/j.landurbplan.2016.11.004

Fischer, L.K., Rodorff, V., Lippe, M. von der, Kowarik, I., 2016. Drivers of biodiversity patterns in parks of a growing South American megacity. Urban Ecosyst 19, 1231–1249. doi:10.1007/s11252-016-0537-1

Huang, G., Cadenasso, M.L., 2016. People, landscape, and urban heat island: dynamics among neighborhood social conditions, land cover and surface temperatures. Landscape Ecol 31, 2507–2515. doi:10.1007/s10980-016-0437-z

Jaganmohan, M., Knapp, S., Buchmann, C.M., Schwarz, N., 2016. The Bigger, the Better? The Influence of Urban Green Space Design on Cooling Effects for Residential Areas. J. Environ. Qual. 45, 134–145. doi:10.2134/jeq2015.01.0062

Lanza, K., Stone Jr., B., 2016. Climate adaptation in cities: What trees are suitable for urban heat management? Landscape and Urban Planning 153, 74–82.

Li, X.-X., Norford, L.K., 2016. Evaluation of cool roof and vegetations in mitigating urban heat island in a tropical city, Singapore. Urban Climate 16, 59–74. doi:10.1016/j.uclim.2015.12.002

Li, X.-X., Norford, L.K., 2016. Evaluation of cool roof and vegetations in mitigating urban heat island in a tropical city, Singapore. Urban Climate 16, 59–74. doi:10.1016/j.uclim.2015.12.002

Vargo, J., Stone, B., Habeeb, D., Liu, P., Russell, A., 2016. The social and spatial distribution of temperature-related health impacts from urban heat island reduction policies. Environmental Science & Policy 66, 366–374. doi:10.1016/j.envsci.2016.08.012

Žuvela-Aloise, M., Koch, R., Buchholz, S., Früh, B., 2016. Modelling the potential of green and blue infrastructure to reduce urban heat load in the city of Vienna. Climatic Change 135, 425–438.

Section II—Ecological Effects of Climate Change

Plants and Climate Change

Leta Ames

As primary producers, plant species form the backbone of most ecosystems on the planet. In addition to primary production, plant species offer a plethora of necessary ecosystem services such as UV protection, air and water purification, and climate stability (Navjot and Ehrlich 2010). These services are essential for all ecosystems to function and provide many necessities and benefits to humans. However, these services are often undervalued and are greatly affected by human activity and climate change (National Research Council 2005).

Since plants are the foundation of many ecosystems, ecological networks—between pollinators, seed dispersers and plants—are more impacted by their extinction than animal extinctions (Schleuning *et al.* 2016). Schleuning *et al.* modeled extinctions using species' climatic niches, defined by temperature and precipitation tolerances. Shifts and disappearances of climatic niches leave few options for species—movement, adaptation, or extinction (Cang *et al.* 2016). However, a plant species' ability to move or adapt to changes in temperature or precipitation does not necessarily guarantee survival. Changes in temperature and precipitation, even those within a species' climatic niche, can affect population dynamics (Petry *et al.* 2016). Additionally, other abiotic factors impact species' ability to resist the effects of climate change. As shown by De Wilde *et al.* (2016), sediment type can impact how well some species of aquatic plants respond to seasonal water level changes. These water level changes may become more extreme in the future due to groundwater pumping for human use and climatic changes (De Wilde *et al.* 2016). Furthermore, as areas become more arid due to climate change, the frequency of fires will increase (Rocha *et al.* 2015).

The frequency and intensity of fires have historically altered ecosystems, even leading to abrupt shifts in vegetation type (Shanahan *et al.* 2016). In turn, these abrupt shifts often cause complete regime shifts—abrupt alterations in the structure and function of an ecosystem (Shanahan *et al.* 2016, Rocha *et al.* 2015). Regime shifts can occur in any ecosystem and are caused by several abiotic and biotic factors such as fertilizer run-off, deforestation, fire frequency, and changes in temperature (Rocha *et al.* 2015). Due to the unique dynamics of every ecosystem, thresholds of regime shifts are difficult to predict and, once established, reversed (Rocha *et al.* 2015, Wernberg *et al.* 2016). Although regime shifts are difficult to predict, developments in models allow for better predictions of future responses to climate change.

Climate change models can be combined with many other factors to predict how plant species will respond to climate change. Schleuning *et al.* (2016) used complex pollinator networks and climatic niche data in conjunction with climate change scenarios to predict extinctions. Genetic history can also be used to help predict future responses to climatic ranges. Phylogenetic trees can be used to understand how organisms have adapted to past change, and in turn how they may be affected by future change. These models have been used to predict the niche shifts of plants, future ranges, and to examine what traits may be the result of evolution or environmental climate change (Cang *et al.* 2016, Rafferty and Nabity, 2016, González-Orozco *et al.* 2016). Although much is still unpredictable, phylogenetic models and climate models provide essential insight into how climate change may affect plant life. These predictions can help inform decisions that can protect ecosystems from collapse.

Plants are the Platform for Ecological Stability

There is a growing need for climate change models that can accurately represent not only the effects on individual species, but also the interactions and compounding effects within ecosystems. These interactions between species form different "mutualistic networks". Schleuning *et al.* (2016) modeled the impact of individual species' responses to climate change in plant-animal mutualistic networks. Specifically, climatic tolerance of 295 plant species, in eight pollinator networks and five seed-disperser networks in unique areas of central Europe were used to understand the relationship between sensitivity to climate change, climatic niche breadth, and biotic specializations.

Distributional and current range data of plant, insect pollinator, and avian seed-dispersal species in the 13 networks were gridded as 3024 squares (50 by 50 km) and were standardized to a European CGRS grid. Climatic ranges of species were determined using temperature and precipitation data for the gridded areas. Although no relationship was found between plants with larger climatic niches and biotic specialization, animals with larger climatic niches tended to have more plant partners in both types of networks. Similarly, animals that were assumed to be more vulnerable to climate change had fewer unique plant partners. The same pattern was not seen in plants.

To model the effects of climate change on extinctions and co-extinctions, Schleuning *et al.* built from previous extinctions simulations. The vulnerability to climate change was predicted by comparing current climate suitability to the suitability in two climate change scenarios. The two scenarios were based off projected average increases of 2.85± 0.62°C and 4.02± 0.80°C, from the Intergovernmental Panel on Climate Change. Extinction was simulated by removing species from the model from highest to lowest vulnerability. Coextinction was modeled by using simulated decreases in the interaction frequencies between partner species. Simulations were created where 25, 50, and 75% decreases in interaction frequency caused an extinction of the partner species. Scenarios with highly specialized partnerships were compared to those with unrestricted partnering of species to account for new partnerships formed as adaptations to climate change.

The simulations demonstrated that animals with more general partnerships were less vulnerable to climate change than specialized animal species. The models also suggest that networks are less able to accommodate loss of plant species than animal species. Plant species' flexibility to animal extinction can be seen in current communities, such as islands. These models suggest that although animal extinctions may slightly impact individual ecosystems, their widespread impact is minimal. Most importantly, these models predict that climate change will exacerbate the effects of plant extinction on biotic networks due to the "cascading effects" of coextinction. These findings suggest that future climate change models should consider biotic interactions when modeling animal responses.

Don't Forget About Dirt: Sediment Type Alters Wetland Plants' Response to Dewatering Events

Wetlands are important ecosystems due to their high biodiversity and productivity, and aquatic plants are vital to their functionality. Human water usage, land usage, and climate change are all threats to the wetland habitat. Water levels change seasonally, however the dewatering experienced in the summer is becoming a major disturbance to aquatic life. Previous studies have investigated the effect of water level changes, including dewatering events (water drainage), on aquatic vegetation. De Wilde *et al.* (2016) build from these studies and compare how different sediment types— coarse, silty, and organic— influence the resistance and resilience of plant communities to changes in summer dewatering.

Over the course of approximately one year, six naturally occurring wetlands along the Ain River in France were surveyed for vegetation type and coverage four times. In each wetland one to two sediment types were sampled to create six sampling units, three solely aquatic and three that experience summer dewatering, for each of the three sediment types. The locations of the samples were chosen based on their similarities in water level and flow to ensure consistency. The coverage of three types of species— hydrophytes, amphiphytes, and helophytes (distinguished by their water tolerance)— were surveyed:

before the dewatering event, at the beginning of the dewatering event, at the end of the event, and two months after the rewetting. To determine the how species abundance changed in each area in response to the dewatering event, they were classified as resident (present before and after), ephemeral (present only during), appeared (present during and after), and disappeared (present only before). The summer dewatering occurred because of a natural decrease in rainfall as well as extraction of groundwater by humans.

To establish a baseline, vegetation abundance was measured before the event; surveys at the beginning and end of the event aimed to measure short- and long-term resistance to dewatering, respectively. Whether the vegetation cover could recover after the dewatering gave insight into the short-term resilience of the community, as measured by the survey two months after the rewetting. Throughout the study water levels were measured and averaged every month. For continuity, these averages were compared to hand measurements taken during the plant surveys.

In the areas that experienced summer dewatering, resistance was significantly lower than that of the solely aquatic areas. Although short- and long-term resistance did not vary significantly in solely aquatic areas, in areas with summer dewatering short-term resistance was lower in the coarse than in silt or organic sediment. During the dewatering event, coarse sediment saw a decrease in resident amphiphytes (plants that grow on the edges of wetlands), which were replaced by helophytes (plants that grow partially submerged). Over the long term, a gradient was formed with organic sediment having the highest resistance, likely due to its ability to retain water, and coarse sediment the lowest. Additionally, resilience was found to increase with resistance, specifically, hydrophytes (plants that grow completely submerged) in silt showed high resilience and could sprout after rewetting.

Leta Ames

Let's Talk About Sex: Population Dynamics may be Modified by Sex-Specific Responses to Climate Change

Many plant species show sex-specific traits, and these differences may impact how climate change affects the species. The differing responses to climate change could impact the ratio between male and female plants, and therefore impact fertility and population dynamics, as well as the range in which the species can survive. Past studies have looked at range shift due to climate change, but Petry *et al.* (2016) look at the pace at which a specific trait (sex distribution) shifts. They studied the response of *Valeriana edulis*, a two-sex herb, to climate change over a 33-year period and across its range from scrubland to mesic alpine tundra (2000–3790 m). Specifically, the sex ratio change, sex-specific mechanisms influencing the change, and the influence of sex ratios on fitness were all assessed over time and across the range.

The data from the study area in the Rocky Mountains in Colorado revealed that for every increase of 100 m in elevation the mean growing season temperature decreased, growing season soil moisture increased, growing season precipitation increased, and delayed the date of snow melt by −0.59°C, 1.09%, 1.5 mm, and 4.1 days, respectively. By finding proportions of flowering plants that are male (operational sex ratios; OSRs) in 31 populations, Petry *et al.* determined that the number of males decreased an average of 0.88% per 100 m increase in elevation. This finding suggested that climate change could produce similar results. In the study area between 1978 and 2014 the mean temperature increased (0.21°C per decade), precipitation decreased (1.91 mm per decade), soil moisture decreased (1.5% per decade) and the snowmelt date occurred sooner (2.9 days per decade). The differences attributed to climate change are equivalent to moving up the mountain at a rate of 175 m per decade.

To understand what causes the OSR shift with elevation and climate, Petry *et al.* compared the life history between the two sexes in four populations. The OSR shifted 22% across a 1167 m elevation change (470–3637 m). Petry *et al.* compared the water use efficiencies (measured carbon assimilated per unit of water transpiration) of eight populations with different OSRs, and found that sex differences in water use efficiency was a good predictor of OSR. In areas with low OSR (female-biased) females had higher water use efficiency, but in high OSR populations the opposite was true. This variation in OSR may also impact the population growth through pollen availability and seedset (female reproductive measurement).

The range of pollen movement was relatively small; males within 10 m of females were responsible for ~90% of pollen. Female seedset was found to increase when more pollen was available. This had an impact to population dynamics because when the male frequencies of low elevations were compared to high elevations, ~50% and 22.8% respectively, median pollen availability dropped by 55% and seedset was reduced from 95 to 76%. Although this species may be insensitive to seed production changes due to its longevity it is important to consider how long-term range shifts due to climate change may impact reproduction. Since the OSR at lower elevations keeps females pollen saturated, the increase in male-bias due to range shift would have little impact on the population, with the exception that the increase in males would replace female individuals. However, for higher elevations the limitations due to pollen availability could dramatically decreased as OSR increases with range shift. Impacts on the OSR may also have a cascading effect since female *V. edulis* are home to more arthropod species at higher densities than males. Although range shifts do not keep pace with climate change, the OSR of *V. edulis* kept up with differences in temperature, precipitation, soil moisture, and snow melt due to climate change.

There is growing research on how climate change will affect biological interactions and the population dynamics of species. Petry *et al.* add to this body of research by providing a better understanding of how population dynamics and sex-specific traits are integral to understand the impacts of climate change.

Leta Ames

Climate Change Resistance Altered by Fire and CO_2 in Tropical Forests

It is well known that fire can play a crucial role in the reproduction and development of plant populations. The availability of water and CO_2 also impact plant growth, especially of larger species. It is believed that the interactions of climate, fire, and CO_2 greatly influence the shift between savanna and tropical forest ecosystems and their permanence thereafter. Previous research has relied on data collected from intact tropical forests, but although useful, these data only provide a snapshot of the impact of CO_2, fire, and climate on these ecosystems. To gain a better understanding of what factors influence tropical ecosystems Shanahan *et al.* (2016) used the concentrations of carbon and hydrogen stable isotopes from sedimentary leaf wax *n*-alkanes ($\delta^{13}C_{wax}$ and δD_{wax}) and the frequency of charcoal layers from sediment obtained from Lake Bosumtwi in Ghana to construct a history of changes in vegetation and hydrology, as well as to estimate the annual fire frequency.

Lake Bosumtwi is in lowland tropical forest, far south of the current savanna-forest edge. The sediment offered a 28,000-year record of hydroclimate and fire frequency of the forest that surrounds the lake. From the sediment record, it was determined that climate, fire frequency, and the dominant ecosystem have shifted significantly in the last 28,000 years. At the beginning of the investigated timeframe (15,000–28,000 year BP) the δD_{wax} values were positive and $\delta^{13}C_{wax}$ values were higher than current levels— indicating a more arid landscape and that drought-tolerant grasses dominated over woody species. The frequency of charcoal layers indicated that fires were more frequent. Nearly annual burns keep woody species from advancing, because stems burned before they were large enough to withstand fires. After 15,001 year BP, more humid conditions with higher precipitation developed, evidenced by decreasing δD_{wax} values. During this time the frequency of fires decreased as well. These declines were gradual, but the $\delta^{13}C_{wax}$ values indicate that the proportion of woody plants changes by ~25% over <150 years— a drastic change. In the remainder of time before present day the proportion of grasses and woody species was variable and mixed. Throughout most of the 28,000 years it appears that precipitation controlled the proportions of grasses and woody species. However, in more recent periods drastic changes in δD_{wax} values were accompanied by minor changes in $\delta^{13}C_{wax}$ values, suggesting that once established, the tropical forest ecosystems could persist in more arid climates. As suggested by other recent studies, the changes in vegetation dominance are the result of interactions between CO_2 concentration, climate, and fire.

Shanahan *et al.* also propose a threshold to explain abrupt vegetation change. The threshold would explain the time lag between changes in hydroclimate and a shift from highly precipitation-dependent vegetation to a system resistant to changes in precipitation. However, not all the changes in vegetation throughout the timeframe support the threshold explanation. Once woody plants dominated, decreased fire frequency, rising atmospheric carbon, and vegetation only responding to precipitation on the centennial to millennial timescale characterized periods without abrupt changes. This is likely because higher CO_2 concentrations promote woody species growth rather than grass growth. Also, woody forests can resist colonization by grass species due to their shaded canopy.

Overall, the impacts of climate, CO_2, and fire shift depending on the role of the other factors. Increased levels of CO_2 were shown to function as a buffer— stabilizing woody forest vegetation— however as the levels rise the stabilizing effects decrease. As the world's climate changes, West Africa could become more arid. If the changes in hydroclimate and fire overcome the stabilizing effects of CO_2, the region could quickly transform from a mosaic of savanna grassland and tropical forest to completely savanna.

Losing Nemo: How Climate Change is Altering Australia's Western Reefs

It is likely that climate change will cause rapid large scale restructuring of ecosystems. Losing species that form the foundation of habitats can drastically change the ecosystem in its entirety, and this is characteristic of regime shifts, due to the distinct ecosystems produced by differing foundational species; these regime shifts can have compounding ecological, economic, and social affects. In oceanic ecosystems, regime shifts of organisms that occupy the lowest water level—the benthic zone—due to the impacts of human activities, have been associated with instability and a decrease of resilience to large environmental changes. Climate change is beginning to have similar effects on ecosystems by altering the distribution of species, and causing interactions between temperate and tropical species in some areas. To better understand the implications of these shifts and the subsequent interactions Wernberg *et al.* (2016) performed surveys of 65 reef ecosystems along the tropical-to-temperate transition zone of the western coast of Australia between 2001 and 2015.

Kelp forests, seaweed, fish, mobile invertebrates, and corals were documented along the approximately 2000 km transition zone. Wernberg *et al.* also measured the temperatures of this region, which borders the Indian Ocean— a "hot spot" with exceptionally rapid rising temperatures. The temperatures along this area have been steadily rising since the 1970s, and in 2011 the region experienced an extreme summer heat wave, one of the three hottest summers of the last 215 years. Before the heat wave, kelp forests dominated over 800 km of the coastline, covering over 70% of shallow rocky reefs. At that time, the difference between the temperate southwest and the more tropical northwest ecosystems were stark, with the midwest more like the south. Two years later, in early 2013, 43% of the kelp forests had been lost, including severe disappearance and loss of forests north of 29°S. Seaweed species had colonized the 370 km^2 vacancy left by the kelp. The establishment of the seaweed corresponded to a shift from temperate to tropical and subtropical species. This shift characterized a "broad-scale community-wide reef transformation". Differences in community structure between northern and southern reefs increased by 91% in seaweeds and 28% in fishes. This restructuring of communities was also seen in invertebrates—coral species increased both in abundance and diversity in the southern midwest. In late 2015, the kelp forests had not returned to their former range or dominance, despite the end of the heat wave.

Previously, after short-term thermal excursions above their upper threshold temperature, kelp forests could slowly recover, but now the kelp forests appear to be suppressed by heavy grazing by subtropical and tropical fish. The biomass of scraping and grazing fishes (groups usually found in coral reefs) increased by 400%, while the seaweed biomass decreased by 80%. In the past five decades, the Indian Ocean has warmed by 0.65°C, and is continuing. The chance that long-term cool conditions needed for kelp populations to recover is decreasing. This large-scale extinction of kelp along most the western Australian coast would decimate the, thousands of temperate endemic species that rely on it, as well as the multi-billion tourism industry.

Watching Grass Grow: Niche Shifts in Grasses are Expected to be Slower than Climate Change

The threat of climate change to biodiversity is becoming increasingly visible. The pace of anthropogenic climate change is faster than previous climatic shifts, making species' responses difficult to predict. When faced with a disappearing or shifting climatic niche, species will likely move up (in elevation or latitude), adapt their climatic niche (through plastic and/or evolutionary responses), or go extinct. Grass species are important to biodiversity: Poaceae (the grass family) consists of over 11,000 species; grasslands cover about a quarter of the land area on earth, and provide habitat for many unique species. Humans also benefit directly from these species—many of them are widely grown crops such as wheat, maize, and rice. Overall, grasses provide 49% of our consumed calories. To understand if these species will be able to adapt to climate change, Cang *et al.* (2016) used phylogenetic trees to determine past adaptation rates and predict future rates.

To determine the past adaptation rates, sister species were compared to their common ancestor by using three time-calibrated phylogenies. The three trees used were each calibrated differently, but were all created using both chloroplast and nuclear sequencing. Climatic data, specifically four factors commonly used for determining climatic niche—mean annual temperature, maximum annual temperature, minimum annual temperature, and mean annual precipitation—were collected from georeferenced localities for each species. Using the mean value of each variable across the localities for each species, the best-fitting model for each sub family for each tree was determined. These models were then used to reconstruct the values for the ancestral species. The niche shift was determined by dividing the absolute difference between the current climatic values of the sister species and the estimated values of their most recent common ancestor. The niche shift rate was then determined using the age of the species. The rate of climate change was determined through four greenhouse gas scenarios and eight standard climate models, each model was used to represent minimum, maximum, and intermediate scenarios. The climate models were then used to project the climate conditions in 2070 for each species. The differences between the projected conditions and "current conditions" (average between 1950 and 2000, with a midpoint at 1975) were divided by 95 years (2070–1975). The average of the values across all localities between all the species was used as the final rate.

By comparing the rates of niche shift to those of projected climate change, Cang *et al.* predict that the rates of climate change will greatly outpace the niche shifts. The exact rate differences varied between climatic variable, the genetic tree used, and the grass species. There are however, many uncertainties that come with predicting species' response to climate change. Although there may be faster rates of niche shift in the future—younger species have faster rates—they would be unprecedented, based on these results. Overall, despite the uncertainty of climate change and niche modeling, these results can give us insight into how climate change may affect biodiversity as well as important food crops.

Phylogenetics for Predicting Phuture Changes

Range models and biome shifts are often used to predict how plants will respond to climate change. However, these predictions don't take into account how plant populations may evolve or how their genetic background may allow them to adapt to the climate change in their current range. To predict future ranges and biodiversity more accurately, González-Orozco *et al.* (2016) created a protocol of how climate change may affect entire clades and landscape level biodiversity by using both species distribution models and measures of evolutionary diversity. Commonly, phylogenetic diversity is used to describe diversity of phylogenetic trees. Phylogenetic diversity uses the sum of branch lengths linking species to represent the shared evolutionary history of species in a region. Although useful, phylogenetic diversity is a limited measure because it doesn't consider rarity. To compensate for this, the phylogenetic endemism—how separated the branches are compared to other areas—was also used. Determining areas of high-biodiversity and rarity can help inform conservation decisions that create and preserve resilient ecosystems. González-Orozco *et al.* tested their approach on the large and diverse clade of eucalypts in Australia.

The clade of eucalypts contains more than 800 species and spans the Australian continent. Although only a few species of eucalypts are endemic outside of Australia, they play important roles in forestry on a global scale. To predict the range shifts of 657 species of eucalyptus, González-Orozco *et al.* used species distribution models in twenty-year increments between 2025 and 2085. Multiple models with and without dispersal as well as considerations of spatial bias and overfitting were used to compensate for known weaknesses, such as climatic adaptability and species interactions, in species distribution models. Previous modeling has predicted that eucalyptus species will be incredibly vulnerable to climate change. This study is no different; the models predict that by 2085, 91% of the 657 species studied will see an average range decrease of 51%. The model also predicted that 2.4% species will become extinct due to the disappearance of their range, but 9% of species are predicted to expand their range. Next, the past and future phylogenetic diversity and endemism were calculated for 25-km by 25-km cells across the continent.

The use of past and future phylogenetic diversity and endemism show how the predicted species losses will be reflected in evolutionary diversity. Losses in phylogenetic diversity are predicted throughout the continent, with an average decrease of 2% by 2085. Also, the turnover rate of phylogenetic diversity is predicted to increase, specifically in the temperate and grassland ecosystems of Southern Australia. The increase in clade change—turnover rate—will be accompanied by increased homogenization of clades. These trends have been seen in Europe and the United States as the results of past climatic events, however the effects of climate change on phylogenetic diversity have not been widely studied for the Southern Hemisphere. As expected, phylogenetic endemism is predicted to decrease due to extinctions and shrinking ranges. It is important to recognize which areas may become endemism and biodiversity hotspots in the future so that they can be preserved.

Budding Innovations in Predicting Phenological Changes Through Phylogenetics

The timing of life-history events, in this case flowering, have shifted in many Northern Hemisphere plant species. These shifts vary in both direction and magnitude between species, and some species have also exhibited changes in their flowering duration—lengthening or shortening. The shifts indicate a changing climate, because flowering is often dependent on temperature and precipitation. Flowering can have huge implications for other species within the plant's community as well as within its own population. Mismatches between pollinators and plant species are unlikely, however they can cause co-extinction when they do occur. Although mismatches don't always occur, flowering shifts can still reshape species interactions, and trait selection. Since the genetics and abiotic factors that alter flowering onset and duration are adequately understood, Rafferty and Nabity (2016) use phylogenetic signal— "the tendency for closely related species to be similar in phenotype"—to understand whether flowering responses to climate change are defined by evolutionary changes or environmental factors.

Previous studies to distinguish between the effects of evolution versus environmental factors have all used data from closely related plants, making the effect difficult to differentiate. To combat this, Rafferty and Nabity used fifteen data sets gathered from studies of flowering onset in plant species in the Northern Hemisphere from many different habitats that spanned at least 20 years. In their analysis, Rafferty and Nabity determined if the flowering onset of species had shifted and determined the slope between the flowering onset and year. To determine the phylogenic signal Rafferty and Nabity compared the slopes that they found with phylogenetic trees, and used statistical tests to find the significance of the fit. Two phylogenetic trees were used to test phylogenetic signal, one was determined using Angiosperm Phylogeny Group 3—a system of angiosperm categorization determined by a group of botanists in 2009—and the other was determined using a molecular phylogeny.

The phylogenetic signal was significant in both flowering time and magnitude on a global scale, indicating that selection molded plant species' response to environmental changes. Additionally, this suggests that the stabilizing selection of flowering time we see currently may shift to directional selection. By using data from multiple communities Rafferty and Nabity also showed how phylogenetic signal can be more effectively used to determine what factors influence phenological response. Rafferty and Nabity suggest that future studies of phylogenetic signal should use data from above the community level and multiple models to more fully understand how climate change will alter ecosystems in the future.

Conclusions

As shown in the research summarized in this chapter, it is difficult to predict the impact of climate change on plants. There is indication that some species can adapt to a changing climate, but, they may not be able to outpace anthropogenic climate change (Rafferty and Nabity 2016, Cang *et al.* 2016). Furthermore, systems are very sensitive to plant species loss and ecosystems that have been pushed past their thresholds may never return to their previous states (Wernberg *et al.* 2016, Schleuning *et al.* 2016). Loss of plant species to climate change could have devastating effects on world biodiversity as well as human life (Navjot and Ehrlich 2010, Cang *et al.* 2016). It is important that we work to further understand the impacts of natural resource use and climate change on plant species so that we can better preserve ecosystems and mitigate the effects of climate change.

References Cited

Cang, F. A. Wilson, A. A. & Wiens, J. J. 2016. Climate change is projected to outpace rates of niche change in grasses. Biology Letters 12, doi: 10.1098.

De Wilde, M. Puijalon, S. & Bornette, G. 2016. Sediment type rules the response of aquatic plant communities to dewatering in wetlands. Journal of Vegetation Science 28, 172–183.

González-Orozco, C. E., Pollock, L. J., Thornhill, A. H. Mishler, B. D., *et al.*, 2016. Phylogenetic approaches reveal biodiversity threats under climate change. Nature Climate Change 6, 1110–1114.

National Research Council, 2005. Valuing ecosystem service: toward better environmental decision-making. The National Academies Press, Washington, DC.

Navjot, S. S., Ehrlich, P. R. (Eds.). 2010. Conservation Biology for All. Oxford University Press, New York.

Petry, W. K. Soule, J. D. Iler, A. M. Chicas-Mosier, A. *et al.* 2016. Sex-specific responses to climate change in plants alter population sex ratio and performance. Science 353, 69–71.

Rocha, J. C. Peterson, G. D. & Biggs, R. 2015. Regime shifts in the Anthropocene: drivers, risks, and resilience. PloS one 10, e0134639.

Schleuning, M. Fründ, J. Schweiger, O. Welk, E. Albrecht, J. Albrecht, M. *et al.* 2016. Ecological networks are more sensitive to plant than to animal extinction under climate change. Nature Communications 7, doi:10.1038.

Shanahan, T. M. Hughen, K. A. McKay, N. P. *et al.* 2016. CO_2 and fire influence tropical ecosystem stability in response to climate change. Scientific reports 6, doi:10.1038.

Wernberg, T. Bennett, S. Babcock, R. C. de Bettignies, T. Cure, K. Depczynski, M. *et al.* 2016. Climate-driven regime shift of a temperate marine ecosystem. Science 353, 169-172.

Effects of Climate Change and Human Activities on Marine Biology

Isabelle Ng

Anthropogenic climate change has been observed to have effects on terrestrial and marine ecosystems, however the impacts of it on marine life remain poorly understood. This chapter discusses recent studies on how human activities and global warming have affected marine organisms and their behavior, demography, ecology, phenology, and physiology.

Atmospheric greenhouse gases (GHG) have been rising over the last 150 years as a result of human activities, and oceans, consequently, absorb one-third of the anthropogenic carbon dioxide (Hoegh-Guldberg & Bruno 2010). Sea surface temperatures have increased by 0.6°C in the last 100 years due to the atmospheric GHGs, and the CO_2 portion of them has also caused ocean acidification, which has been shown to affect marine ecosystems based on calcification such as coral reefs, as well as other organisms' metabolic systems (Hoegh-Guldberg & Bruno).

Heating also impacts the behavior of ocean currents which strongly affects ocean dynamics. It could also affect multi-annual cycles such as the El Niño–Southern Oscillation (ENSO), North Atlantic Oscillation (NAO), and Pacific Decadal Oscillation (PDO) (Hoegh-Guldberg & Bruno).

The largest observable effects of global warming are most apparent in polar oceans. Temperature increases there are twice the global average, which has led to massive decreases in sea ice and ice sheets (Hoegh-Guldberg & Bruno). Higher rates of Arctic sea ice melt and earlier ice breakup has contributed to increasing sea levels.

In this chapter, I will discuss how global climate change impacts biological processes of larger marine organisms such as seals, whales, dugongs, sea turtles, and sharks. Temperature shifts have an inherent effect on energy distribution, which then lead to changes in an organism's enzymatic functions, diffusion rates, and membrane transport, which are all fundamental biological processes (Hoegh-Guldberg & Bruno). Increases in temperature may also increase metabolic rates, which would then impact an organism's life history traits and population growth, with eventual effects on the entire ecosystem (Hoegh-Guldberg & Bruno). Some organisms are able to physiologically adapt to temperature changes that are not too far off the global average or their optimal range (Hoegh-Guldberg & Bruno). However, beyond this range, organisms may fail to adapt and thus face a higher chance of mortality, reduced fitness, population decline, or possibly extinction (Hoegh-Guldberg & Bruno). An organism's phenology may also change as a result of global climate change (Walther *et al.* 2002). Whales, for example, may change their migration patterns as a result of warmer sea surface temperatures or earlier sea ice melt.

Temperature fluctuations can also lead to shifts in fundamental biological processes such phytoplankton productivity. Changes in phytoplankton distribution and abundance will affect ocean's primary productivity and the organisms that rely on them.

This chapter also discusses how environmental conditions such as episodic climatic events, shifts in sea ice formation, food availability, and climate variables affect marine organisms' demographics, ecology, phenology, and physiology. These are observed through variables such as migration timing, reproduction, metabolism, and behavior.

Environmental Variables and Episodic Events Cause Demographic, Ecological, and Physiological Changes in Hudson Bay Ringed Seals

Shifts in environmental conditions can lead to demographic, ecological, and physiological changes in organisms. Endemic Arctic species in particular, face threats from sea ice changes as a result of global warming. Researchers from Canada (Ferguson *et al.* 2017) focused their study on the southern distributed ringed seals in the Hudson Bay. Ringed seals require sea ice during the spring season when they undergo reproduction and molting as well as face higher predation risks. During the summers, ringed seals rely on the open water to forage enough to gather energy for the winter. As a result, changes in environmental conditions can cause shifts in the phenology of ringed seals.

The Hudson Bay undergoes a cycle of seasonal ice formation and loss. The researchers wanted to look at how various environmental features such as sea ice and climate indices compared to the seal's biological variables from 2003 to 2013. They looked at seal body condition in terms of fat percentage, reproduction, and recruitment from hunter harvest data, as well as stress measured via cortisol. The researchers gathered these data by testing seals that were caught through Inuit subsistence hunting.

They observed that over the study period, sea ice breakup occurred earlier in the spring while formation occurred later in autumn. Serious declines in sea ice led to gradual shifts in environmental conditions and climatic events. The data showed that blubber mass significantly decreased over time, meaning ringed seal body conditions deteriorated. A relationship between body condition and open water period was found. The longer the open water period, the more ringed seal body conditions decreased. Pup survival also declined slightly over the 10 year period. Furthermore, cortisol levels were observed to have significantly increased. High stress levels were recorded in 2010, which was the same year of the climatic event where ice breakup occurred unusually early. The following year, low ovulation rates were found and consequently, a lower count of seal pups. However, the ringed seals displayed significant environmental plasticity years after the event by returning to high ovulation levels and fat percentages.

Another reason for the observed lower body fat percentage could be diet shifts. Since 2000, there was a decrease in abundance of Arctic cod in the Hudson Bay which forced the seals to consume sub-Arctic capelin and sand lance instead. Arctic cod have the highest energy content relative to other species the seals eat, so this diet change may have severely impacted the seals' body conditions. Shifts in the type of prey available could be due to higher sea surface temperatures from global warming. A warmer ocean could have also caused increased competition by temperate species, higher predation from those that were previously blocked by sea ice, and a greater occurrence of disease due to higher stress levels.

The observed decline of ringed seals after the 2010 climatic event may have also been due to sea ice loss which left them with no resting platform to molt, thus making them more susceptible to diseases. Furthermore, in the autumn, seals are at their maximum blubber fatness, which led to higher levels of lethargy and thus increased predation risk to polar bears.

This research sheds light on the fact that endemic Arctic species such as the ringed seal are already undergoing demographic, physiological, and ecological changes due to global warming.

Changes In Annual Sea Ice Formation Are Causing Migration Timing Shifts of Pacific Arctic Beluga Whales

Climate change has been observed to cause shifts in seasonal environmental gradients, most prominently in the Arctic. Marine species such as Pacific Arctic beluga whales have migration timing that is temporally matched to seasonal sea ice cover, which means their migration patterns may shift as a result of climate change. Researchers wanted to find out whether the sympatric Eastern Chukchi Sea and Eastern Beaufort Sea beluga populations experienced changes in their migration timing due to delayed sea ice freeze-up.

Sea ice changes have become more frequent and intense since the 1990s. Both of the beluga populations undergo sea ice conditions that could jeopardize their ability to safely migrate from their winter regions in the northern Bering Sea to their summer foraging areas in the Pacific Arctic. The researchers predicted that assuming sea ice significantly affected beluga migration, they would observe a shift in later migration timing as sea ice freeze-up occurs later over time.

The researchers used two methods – satellite telemetry and passive acoustics – to observe both populations. Both Chukchi and Beaufort beluga whales were tagged with satellite-linked data recorders in order to study their locations. Hydrophones were additionally deployed on moorings to record vocalizations made by both populations when passing by.

The data revealed that the Chukchi population migrated significantly later from the southern Chukchi Sea in the autumn months of 2007–2012 in comparison to earlier years. Migration timing for Chukchi belugas was consistently shifted in areas from which the whales were departing, which could mean that foraging regions are not easily accessible due to the formation of autumn fast ice, which has been delayed since 2000. These findings suggest that the Chukchi populations are able to phenologically adapt to delays in sea ice freeze-up timing.

The Beaufort population on the other hand, had no observed shifts in migration patterns. This is possibly because Beaufort beluga whales are not able to sense sea ice changes in the western region of the Chukchi Sea as well as the eastern side, where the Chukchi belugas reside.

A fault of this study, however, was its small sample sizes. As such, conclusions about phenological adaptations to sea ice changes cannot be drawn with confidence. While a larger sample size and longer time frame is desirable, this study was still able to reveal that phenological responses to environmental changes could be different between sympatric beluga whale populations.

Isabelle Ng

The Impact of Climate and Food Availability on Southern Right Whale Reproductive Success

Researchers from Brazil and Australia sought to explore how food availability, climate and oceanographic factors impact the reproductive success of the southern right whale population. The purpose of the study was to investigate whether krill density near South Georgia is related to calf production in the whale's main breeding ground off Brazil.

Annual aerial surveys were conducted in heavily concentrated breeding locations of southern right whales from 1997 to 2013. Seyboth *et al.* (2016) used the number of calves observed as a measure of reproductive success, correcting for an underlying positive trend of calf counts as the population is still recovering from the 1973 ban of commercial whaling in Brazil. In order to make the correction, the researchers detrended correlation analyses data. Effects of climate and oceanographic factors were derived from climate indices such as the Oceanic Nino Index (ONI), Antarctic Oscillation (AAO), SST anomalies around South Georgia (SSTSG) as well as sea ice anomalies.

The results of the study showed that there was a relationship between the number of calves observed in southern Brazil and the abundance of krill near South Georgia, which is a vital feeding area for the southern right whale population. This supports the authors' initial hypothesis that the southern right whale's reproductive success is impacted by the amount of food available during their early gestation period. The amount of nutrition obtained by the whales can influence various stages of their reproductive cycle, however, the feeding conditions during gestation appears to have a more significant impact on calving rates.

Krill densities were also found to be impacted by climatic factors. The krill population in South Georgia relies on input by the Antarctic Circumpolar Current, which means that krill density can be impacted by shifts in oceanic circulation. Additionally, elevated sea surface temperatures in South Georgia led to a decline in krill recruitment that same year, which subsequently decreased krill biomass the following summer. These declines in krill density were significantly correlated with calf production in Brazil months later.

A former study which measured the impact of climate change scenarios on krill density discovered that a 1°C increase over the next 100 years would lead to a 95% decline in krill biomass over a span of 50 to 60 years. This massive reduction could severely impact southern right whale population dynamics.

Effects of Climatic Drivers on Dugong Calf Production vary Geographically Along the Eastern Queensland Coast, Australia

Climate variables have been observed to have an effect on dugong demographic parameters in Queensland, Australia, where dugongs occur at high densities. Mortality, mass strandings, weight loss, and delayed reproduction occur in dugong populations as a direct result of extreme weather events. Dugongs are herbivorous and rely heavily on seagrass as their food source. Their responses to climate drivers are most probably due to the immense loss of seagrass communities.

During the summer of 2010/11, the Queensland coast was affected by severe weather events, including floods and tropical cyclones which resulted in years of seagrass deterioration. Researchers from James Cook University studied the effects of four climate variables – rainfall, the Southern Oscillation El Niño Index, sea surface temperature, and frequency of tropical cyclones – on dugong calf production in various sub-regions along the eastern coast of Queensland. Aerial surveys were conducted since the mid 1970s to observe calf sightings in five sub-regions. The number of calves seen during a survey was reflective of the frequency of births and calf survivorship, which then exposed the impact of climate drivers on female fecundity and mortality.

The results revealed that the number of calves varied among the five sub-regions. One common theme was that the proportion of calves was negatively correlated with higher average rainfall and cyclones, which are direct consequences of La Niña events. Such episodes were presumed to have caused seagrass loss. Sea surface temperature changes as a result of El Niño events were also found to cause a decline in calf production. Sea surface temperature changes may also severely impact the survival of Eastern Australian sub-tropical and tropical seagrass species. Varying seagrass community composition and ability to tolerate stressors may explain why there were different responses among the sub-regions.

The researchers also observed that the impacts of climate variables were consistently lagged. This means that the proportion of dugong calves is more heavily influenced by the state of the mother before and during pregnancy and lactation rather than by current extreme weather events.

Climate variables not only impact the number of calves, but also the survival of adult dugongs. The study revealed that adult mortality was caused by smaller scale climate changes such as increased freshwater input and lowered air temperatures which were associated with La Niña events. The authors argue that future conservation measures should take into account these indirect effects by implementing habitat protection and coastal management. Successful management is crucial as climate change may further exacerbate the effects of these stressors on seagrass communities and dugong populations.

Climate Change Increases Frequency of Female Hatchlings in North Carolina Loggerhead Sea Turtle Population

Sea turtles' sex determination is affected by the temperature range they are exposed to during incubation. A greater ratio of females is observed at high temperatures while males are produced at lower ones. The optimal temperature for an even ratio of males to females is 29°C, and higher surface temperatures will lead to a greater proportion of female hatchlings.

Researchers from the University of North Carolina investigated the effects of temperature and precipitation on the incubation duration, sex ratio, and hatchling success in a loggerhead nesting beach on Bald Head Island, North Carolina. The goal of their study was to assess whether environmental changes from 1991 to 2015 had ramifications on loggerhead hatchling sex ratio and survival. Loggerheads are a particularly important species to study because they are listed as Threatened by the IUCN and their populations are already highly female-biased. Further impacts from climate change could thus place loggerheads at a risk of extinction.

Increased temperature is known to significantly decrease incubation duration. The researchers found that over the last 25 years, temperature significantly increased during the nesting season, which led to shorter incubation times. As a result of these changes, there was an increase of 55% to 88% ratio of female hatchlings. Previously, the loggerhead population at Bald Head Island has had a balanced sex ratio and oftentimes provided males to mate with the female-skewed populations in the south. The authors argue that management of the Bald Head Island loggerheads should be a priority in order to preserve this vital nesting habitat and secure the survival of it's surrounding populations.

While temperature did not impact hatchling survival, precipitation did. Higher rainfall meant higher hatchling mortality due to suffocation. This finding may threaten sea turtles in the future, as climate change is projected to cause more frequent and heavy rainfall.

Effects of Climate Change on Nesting and Internesting Leatherback Seaturtle Core Temperatures

Researchers at the University of Wisconsin-Madison used novel model-based methods to analyze how climate change will impact nesting and internesting in leatherback sea turtles, with particular emphasis on their temperature regulation. Leatherback sea turtles are especially difficult to study because their large body size prevents them from easily losing excess heat, thus increasing their core temperature. As a result of this trait, scientists have difficulty figuring out what leatherback core body temperatures should be. Leatherbacks also have a very intricate life history, with the largest range of all reptiles. Leatherbacks spend majority of their life in the ocean but nest on beaches. Consequently, when researching leatherbacks it is necessary to consider both their marine and terrestrial distribution. The researchers counteracted these difficulties by using an animated 3D computational fluid dynamics (CFD) simulation, and internal heat transfer and ecological niche models.

The modelling was divided into three parts for both the internesting and nesting phases: the CFD simulation step, the heat transfer simulation step, and the niche model step. Five sizes of leatherback turtles were designed with a 3D software for the CFD step. Long narrow, long wide, short narrow, short wide and average ratio were the sizes chosen based on previous research on sampled leatherbacks. Heat transfer and power needed for a set swimming speed were calculated for each size model. The next step, the heat transfer simulation, took the data from the CFD to determine core temperatures under different environmental conditions. The final step, the marine niche model step, combined the water temperature and core temperature data with four climate models to test how leatherback core temperatures would change under various scenarios.

The model showed that average sized leatherbacks with large nesting populations in French Guiana and Suriname (South America), Gabon and Congo (Central Africa), and West Papua (Indonesia) will have higher internesting core temperatures. West Papua (WP) showed the highest increases in core temperature, which implies that populations in Southeast Asia will be the most impacted by global warming.

Leatherback populations in WP can physiologically combat higher temperatures by changing their nesting time and locations. The model showed that shifting the nesting time was not so effective for the WP populations, while location change was slightly helpful. Having a smaller body size is advantageous as they are able to keep their core temperature lower, but shifts to a smaller body size could also lead to a decline in generation time. Other physiological adaptations to higher temperatures not considered in this model could include a reduction in swimming speed or increase of resting time. These adaptations however, may have implications on their fitness.

This study revealed how physiological data can be used to predict how global warming affects an organism's future fundamental niche and range. This information will be beneficial to conservation and management around nesting beaches, especially with the more sensitive populations in Southeast Asia.

Combined Effects of Global Warming and Ocean Acidification on Juvenile Bamboo Sharks in the Indo-West Pacific

Global warming and ocean acidification have varying effects on marine organisms. This study in particular, shows how the combination of effects from warming and acidification impacts the digestive enzyme activity of small bottom-dwelling juvenile bamboo sharks in the coastal Indo-West Pacific. Sharks aren't able to quickly adapt to environmental changes due to their low fecundities, varying lengths of gestation, low population growth rates, and long life span. Furthermore, shark's digestive systems work slowly, which in turn limits their growth and reproduction.

Rosa *et al.* (2016) collected 60 recently spawned embryos of bamboo sharks from Cebu, Philippines, then incubated and acclimated them at varying temperatures and acidities. The embryos were incubated at (1) controlled temperature and pH values, (2) a temperature control and acidification scenario, (3) a pH control and warming scenario, and (4) a warming and acidification scenario. The warming scenario was ocean warming expected in 2100, which is 4°C above average ambient temperature (26°C) in the Indo-West Pacific region. The acidification scenario was set at pH 7.5 relative to the pH 8 control. Thirty days after hatching, two groups of digestive enzymes were measured: trypsin, a pancreatic enzyme and alkaline phosphatase (ALP), an intestinal enzyme. These enzymes regulate metabolism and thus are frequently used in studies as markers of fish development, condition, and physiology.

The sharks showed a significant increase in trypsin levels during the warming scenario and a decrease during acidification scenarios. This meant that acidification and warming were acting in opposition. Similar trends were found for alkaline phosphatase levels. The lowest values of trypsin were observed during the temperature control and acidification scenario. This decrease in digestive trypsin enzyme levels due to acidification could slow shark digestion even further and thus decrease metabolism and fitness.

Tourism Causing Behavioral Changes of Whale Sharks in Western Australia

Western Australia's Ningaloo Marine Park (NMP) is one of the few locations in the world where whale sharks are known to aggregate, which makes it a popular destination for nature-seeking tourists. Tourism levels are high between March and July, when whale sharks aggregate in high-nutrient waters. While tourism may benefit from these aggregations, the whale shark is threatened and listed as Vulnerable by the IUCN, most likely a result of human impacts such as tourism. The whale shark tourism industry is managed by the Department of Parks and Wildlife under a species and management program, which is supposed to exercise "sustainable best practices" through a code of conduct.

Over a two-year period, researchers conducted aerial surveys to evaluate the impact of tourism vessels and snorkelers on whale shark movements. Previous studies on the whale shark tourism industry in the NMP were conducted in water, which may have affected the whale sharks' behavior at the time of the study.

This study revealed that the close proximity of vessels and snorkelers to whale sharks caused a higher frequency of directional movement, with twice as many changes in direction than otherwise. Whale sharks dive and resurface to regulate their internal temperature, which means tourist and vessel presence could potentially disrupt this vital process. Moreover, whale sharks may become habituated to humans and vessels, putting the already vulnerable species at a greater risk. There is also the danger of physical collisions between whale sharks and vessels.

While the implications of the whale shark tourism industry are now known, the costs have yet to be measured. Other factors such as resultant noise from vessels and tourists should also be considered in future research. However, management decisions and changes to the NMP program's code of conduct can now be made with the consideration of these findings.

When tourism and nature interact, it is often found that there are controversies and environmental implications. Human related impacts such as the vessel movement are observed to be the main threat to whale sharks in Western Australia. It is important to understand the effects of such disturbances, evaluate the current policies in place and make the necessary changes to minimize these threats.

Conclusions

Around the world, and most prominently in the Arctic, marine organisms have been forced to undergo shifts in their ecological niches due to global warming. This chapter discussed recent research showing how marine organisms must overcome such changes through phenological or physiological adaptations. However, even if some marine organisms are able to adapt, it does not necessarily mean they are able to adapt sufficiently for the population to survive. For example, in the case of the West Papua leatherback sea turtles, shifting of nesting time was not at all effective and location change only helped slightly. Furthermore, any time an organism has to adapt to a changing climate, it requires more energy to do so, which in turn may impact their fitness. As observed from these examples, global climate change may lead marine organisms to shift their biological functions but these adaptations may in turn require more energy output and thus cause reductions in fitness.

It is vital that scientists continue observing the effects of global warming on marine ecosystems, which are not as frequently studied relative to terrestrial ecosystems. Conservationists must work in coordination with such scientists in order to create effective management plans for endangered marine life and their habitats. Furthermore, managers and policy makers must make it a priority to understand how specific risks impact marine organisms and their environments, and work to reduce and prevent such risks. More importantly, increased effort must be made to reduce greenhouse gas emissions. As long as our Earth's oceans remain, they will continue to absorb heat produced by anthropogenic activities and marine organisms will also suffer as a result.

References Cited

Dudley, P.N., Bonazza, R. and Porter, W.P., 2016. Climate change impacts on nesting and internesting leatherback sea turtles using 3D animated computational fluid dynamics and finite volume heat transfer. *Ecological Modelling, 320*, pp.231-240.

Ferguson SH, Young BG, Yurkowski DJ, Anderson R, Willing C, Nielsen O. 2017. Demographic, ecological, and physiological responses of ringed seals to an abrupt decline in sea ice availability. *PeerJ* 5:e2957

Fuentes, M. M., Delean, S., Grayson, J., Lavender, S., Logan, M., & Marsh, H. (2016). Spatial and temporal variation in the effects of climatic variables on dugong calf production. *PloS one, 11*(6), e0155675.

Hauser, D. D., Laidre, K. L., Stafford, K. M., Stern, H. L., Suydam, R. S., & Richard, P. R. (2016). Decadal shifts in autumn migration timing by Pacific Arctic beluga whales are related to delayed annual sea ice formation. *Global Change Biology*.

Hoegh-Guldberg, O., & Bruno, J. F. (2010). The impact of climate change on the world's marine ecosystems. *Science, 328*(5985), 1523-1528.

Raudino H, Rob D, Barnes P, Mau R, Wilson E, Gardner S, Waples K (2016) Whale shark behavioural responses to tourism interactions in Ningaloo Marine Park and implications for future management. *Conservation Science Western Australia* 10: 2 [online].

Reneker, J. L. and Kamel, S. J. 2016, Climate change increases the production of female hatchlings at a northern sea turtle rookery. Ecology, 97: 3257–3264. doi:10.1002/ecy.1603

Rosa, R., Pimentel, M., Galan, J.G., Baptista, M., Lopes, V.M., Couto, A., Guerreiro, M., Sampaio, E., Castro, J., Santos, C. and Calado, R., 2016. Deficit in digestive capabilities of bamboo shark early stages under climate change. *Marine biology*, *163*(3), pp.1-5.

Seyboth, E., Groch, K.R., Dalla Rosa, L., Reid, K., Flores, P.A. and Secchi, E.R., 2016. Southern Right Whale (Eubalaena australis) Reproductive Success is Influenced by Krill (Euphausia superba) Density and Climate. *Scientific Reports*, *6*.

Walther, G. R., Post, E., Convey, P., Menzel, A., Parmesan, C., Beebee, T. J., ... & Bairlein, F. (2002). Ecological responses to recent climate change. *Nature*, *416*(6879), 389-395.

Is Climate Change Causing Coral Reefs to Disappear for Good?

Natalie Ireland

Coral reefs provide habitat for many marine animals and coastal protection from waves. Coral reefs around the world are being threatened by rising ocean temperatures, increasing levels of CO_2, and ocean acidification. Anthropogenic CO_2 emissions are the driving force behind these three factors that are putting large amounts of stress on vulnerable coral reefs. Large amounts of CO_2 are released into the atmosphere, which leads to more CO_2 being stored in our oceans. Because of this, corals are more susceptible to massive bleaching events, loss of zooxanthallae, and loss of demersal zooplankton. Human activities are also causing coral reefs to disappear at an alarming rate. Human activities such as overfishing, sedimentation, and nutrient enrichment are allowing algae to grow and thrive in places where coral once lived. These human activities coupled with natural phenomena such as widespread bleaching events, cyclones, or corallivore predation is leading to the sharp decline in live coral cover on coral reefs around the world. Coral reefs are home to numerous types of marine life, and the degradation of corals is threatening many marine species. Fish and other marine animals that live within corals are losing habitat and food sources as corals disappear.

Scientists are now researching new strategies, techniques, and possible refuges for corals suffering from high temperatures and acidic waters. Moderately turbid waters, small herbivores, and mangroves and seagrass habitats may help save corals from complete destruction. Moderately turbid waters shade coral by suspending small particles in the water column. Small herbivores eat algae growing on damaged corals. This allows damaged corals to grow back in previously damaged areas. Mangrove and seagrass habitats buffer corals from lowering pH levels. It is important to research corals that have adapted to the changing ocean conditions; these corals can give insight into how to save other corals from completely dying off. Coral reefs are important marine ecosystems, and if coral reefs die off, many other marine animals will also die soon after.

Ocean Acidification and High Temperatures Threaten Coral Species off the Coast of Hawaii

Climate change has caused an increase in ocean temperatures and an increase in pCO_2 levels, leading to a decline in coral reef ecosystems. Ocean acidity has increased by about 25% since preindustrial times, and coral calcification rates are expected to decrease by 40% by the end of the century due to decreasing ocean pH. Rising ocean temperatures are expected to increase thermal bleaching of corals, reduce calcification, and increase coral mortality. Almost every reef around the world has experienced increased stress and mortality due to climate change. Increased sea surface temperatures and increased levels of CO_2 have been shown to independently cause coral decline, however it has not been tested if these factors cause a greater decline in coral when working together. Bahr *et al.* (2016) set out to discover if multiple stressors working in unison have a larger negative impact on growth and calcification of corals than the stressors working alone. They tested increased sea surface temperatures, and increased levels of pCO_2 on five different coral species (*Porites compressa, Pocillopora damicornus, Fungia scutaria, Montipora capitata,* and *Leptastrea purpurea*) off the coast of Hawaii.

They found that increased sea surface temperatures had a negative effect on calcification rates and increased the amount of exposed dead skeleton on three coral species (*P. damicornus, M. capitata,* and *P. compressa*). Increased temperatures had no effect on calcification rates or percentage of exposed dead skeleton on two species of coral (*F. scutaria,* and *L. purpurea*). Increased levels of pCO_2 had a negative impact on calcification rates of two coral species (*P. damicornis*, and *P. compressa*), and had no effect on three coral species (*F. scutaria*, *L. purpurea*, and *M. capitata*). Increased pCO_2 levels increased the amount of exposed dead skeleton on the coral species *P. compressa*. *P. compressa* was also the only coral species that had an increase in percentage of exposed dead skeleton when sea temperatures increased and levels of pCO_2 increased. *P. damicornis* was the only coral species that had a decrease in the rate of calcification when both temperature and pCO_2 levels increased. Responses of calcifying organisms to both temperature changes and changes in levels of pCO_2 are variable and species-specific. This study showed that rising temperatures have a greater effect on coral reef growth and calcification than changing levels of pCO_2, and both factors combined had little or no additive effect on the five coral species tested.

Effects of Ocean Acidification on Demersal Zooplankton Living on Coral Reefs

The amount of CO_2 in the ocean is increasing due to increased concentrations in the atmosphere, causing the ocean to warm and become more acidic, and making it difficult for zooplankton to survive. Zooplankton play an important role in many marine ecosystems; they are food for many planktivores, they support bacterial production by excreting nitrogen and phosphorus, and they act as a sink for CO_2. Zooplankton living on coral reefs are demersal, meaning the organisms live on substrata during the day and reside in the water column at night. A loss of demersal zooplankton could be disastrous for coral reefs. Corals rely on demersal zooplankton for a source of essential nutrients they cannot acquire themselves. Demersal zooplankton are especially important to corals as the threat of ocean acidification increases; corals will need more nutrients to meet the increased energy demands for calcification.

Smith *et al.* (2016) found a loss in zooplankton biomass when comparing control coral reef sites and a coral reef sites with a high amount of CO_2. There was a reduction in taxa abundance at the high CO_2 sites compared to the control sites. Zooplankton species were also not able to proliferate in the high CO_2 sites. It was observed that the high CO_2 reef sites studied were composed of massive bouldering corals, while control sites were composed of structurally complex corals. Diminished coral reef complexity due to ocean acidification negatively impacts organisms, such as zooplankton, that rely on coral reefs for habitat. Massive bouldering corals at high CO_2 sites caused the reduction of zooplankton taxa at these sites. Less complex coral structures reduced the amount of habitat for zooplankton and other small invertebrates relying on branching coral for refuge. This study showed that coral reefs may be in more danger than expected from ocean acidification because rising amounts of CO_2 are killing the basis of their food webs.

Climate Change Could Cause Intense Cyclones, Which Could Spell Disaster for Coral Reefs

Climate Change has many negative effects on coral reefs, including creating widespread bleaching events. The most damaging phenomenon climate change has created is more intense cyclones that destroy coral structures and displace dependent species. Cyclones occur naturally, but the destructiveness and intensity of cyclones are increasing due to warming ocean temperatures and acidic waters. Usually, cyclones hit only part of a coral reef leaving some of it intact. Undamaged communities are able to replenish the damaged areas within a few decades, however, this reduces diversity and taxonomic richness of coral communities. Global climate projections predict that there will be an increase in the frequency of the most intense cyclones as climate change increases. Cheal *et al.* (2017) set out to discover how coral reef communities would be affected by the predicted changes in cyclone activity.

The authors studied how three category 5 cyclones affected the Great Barrier Reef. They found that after the cyclones hit, coral cover declined and severe destruction was observed, compared to stable coral cover at the unaffected coral reefs. Reef fish communities also suffered greatly from the intense cyclones. Directly after the cyclones, fish species diversity dropped to the lowest level recorded, and local extinctions of fish increased to 23 species. Fish species richness in the unaffected coral regions also decreased, but at a much slower rate than in the affected areas. Smaller-bodied and coral-dependent fish suffered the most after the cyclones, and larger species also experienced high losses. Large herbivore species, however, increased in abundance after the cyclones.

The authors concluded that average cyclone frequency in the Great Barrier Reef will not increase under climate change, but the frequency of intense cyclones will increase due to climate change. Intense cyclones are predicted to increase to one every 25 years by 2100. Predicted increases in cyclone intensity will reduce biodiversity and increase the vulnerability of coral reefs to degradation by bleaching.

How Sedimentation, Nutrient Enrichment, and Overfishing Impact a Coral Reef Ecosystem Immediately Following a Disturbance

Coral reefs are regularly disturbed by natural phenomena such as bleaching, storms, and outbreaks of predators, such as the coral-livorous sea star *Ancanthaster planci.* Corallivores are animals that eat coral polyps. Coral reef ecosystems are resilient, and are often able to recover from large-scale disturbances quickly. However, anthropogenic stressors such as overfishing, nutrient enrichment, and sedimentation can prevent coral reefs from recovering. Nutrient enrichment, caused by terrestrial runoff, creates the perfect environment for benthic algae to grow on disturbed and broken coral reefs. Overfishing, working in tandem with nutrient enrichment, causes an overgrowth of algae if there are not enough fish to graze it, and the successive degradation of the reef. Sedimentation is another side effect of terrestrial runoff. Sedimentation buries corals, which blocks light from reaching them and potentially stops coral recovery. However, sedimentation, when not paired with any other stressor, can also stop the growth of algae by burying surfaces that algae would grow on. Gil *et al.* (2016) set out to test the interactive effects that overfishing, sedimentation, and nutrient enrichment have on coral reefs in French

Polynesia. They hypothesized that these anthropogenic disturbances, when working interactively, will negatively impact corals, while promoting algal cover.

To test their hypotheses, the authors used two different corals, *Acropora pulchra,* a thin-branching coral, and *Porites rus,* a bulbous coral to test the effects of these anthropogenic stressors on corals with different morphologies and ecological functions. This study found that nutrient enrichment and overfishing together resulted in the largest increase in algal turf, while sedimentation, paired with either nutrient enrichment or overfishing, decreased algal turf growth; macroalgae biomass growth was significant when overfishing occurred, but insignificant when nutrient enrichment or sedimentation occurred. The coral species *A. pulchra* was unable to be tested for survival because during the experiment, 100% of the tested corals were eaten by corallivores. The coral species *P. rus* was tested for percent survival in each anthropogenic stressor scenario. Sedimentation and overfishing working together did not hinder the coral from recovering. This scenario had the most live coral cover by the end of the experiment, while the other scenarios were statistically insignificant. This study was important because it showed that different coral morphologies act differently when under the same anthropogenic factors.

This study showed that while nutrient enrichment and overfishing can cause both algal turf and macroalgae growth in disturbed coral areas, high amounts of sedimentation can prevent algal growth in most coral reefs. Coastal ecosystem managers can use this information to better understand and control benthic algal blooms. Controlling terrestrial activities that create runoff and controlling overfishing could be the answer to suppressing algal blooms that are preventing coral from recovering after an initial natural disaster.

Changes in the Diet of Juvenile Coral Trout After Coral Reef Degradation

Extreme ocean environmental disturbances are becoming more common as sea levels rise, the ocean becomes more acidic, and global temperatures increase. These environmental disturbances can have harmful impacts on coral reefs and the animals that live within and around them. Coral reefs provide important habitats for many marine animals such as crustaceans, invertebrates, and coral fish. When reef habitat quality is diminished after an environmental disturbance, habitat and food availability for coral reef fishes can change. Most coral reef fish are able to quickly adapt and shift their diet based on food availability, however their susceptibility to environmental changes differs throughout their life cycle. In this study Wen *et al.* (2016) examine how the dietary composition of coral trout, *Plectropomus maculatus*, changes after a reef disturbance. The authors studied newly settled recruits and juveniles and compared the ability of these fish to shift their diets based on food availability.

Wen *et al.* collected coral trout around Keppel Island in Southeast Queensland before and after a 2011 river flood plume disturbance. After the disturbance, live coral cover and the number of coral trout in both life stages decreased. The authors also found that dietary shifts occurred in both the newly settled recruits and the juveniles after the flood disturbance and the loss of live coral cover. Before the disturbance, newly settled coral trout ate small crustaceans, but after the flood, plankton was found primarily in the stomachs of recruits. Dietary shifts in juvenile coral trout also occurred. Before the flood, juveniles mostly consumed prey fish, however after the flood, larger amounts of crustaceans than prey fish were found in the guts of juveniles. The type of prey fish and crustaceans consumed also changed following the disturbance. Coral trout were able to shift their diet to whatever fish were available when foraging in a degraded coral reef. This research suggests that juvenile coral trout have more food selectivity when foraging around a live reef, and are able to quickly adapt their feeding habits based on food availability, however there were fewer coral trout found around the reef after the disturbance. It is still unknown if there are any consequences to the fitness of coral trout when change in diet occurs.

Natalie Ireland

The Important Role Small Herbivores Play on Degraded Coral Reefs

Biodiversity is constantly being altered by anthropogenic and natural variants. Due to ocean acidification, and rising ocean temperatures, coral reef systems have degraded, and algae has come to dominate some of these systems. Macroalgae are aggressive and quickly colonize areas where coral has been degraded, and heavy algae cover of dead coral substrates prevents recovery of dead coral communities. A study conducted by Kuempel and Altieri (2017) set out to discover how coral reefs adapt to changing environments and how individual species living along the reefs promote resilience. The presence of herbivores, such as parrotfish, sea urchins, and other small grazing fish around degraded coral reefs likely halts the shift from coral-dominated areas to algae-dominated areas. Understanding the rate of recovery for coral reef dynamics can help scientists predict future coral resiliency and aid conservation efforts.

Kuempel and Altieri studied coral reefs on the Caribbean coast of Panama after a recent hypoxic event killed over 90% of coral on some reefs in that area. They chose to study this area because it has high anthropogenic stress, increasing the chance of a higher rate of algal dominance after coral disturbances. Using field surveys, herbivore manipulation, caging, and algal transplant, Kuempel and Altieri were able to study the relationships between herbivore populations, pressures that herbivores face, and grazing importance in relation to other algal mitigating factors.

This study found that there was no correlation between mass coral reef deaths and high rates of macroalgae cover. A large number of herbivores, mostly small grazing fish and invertebrates, around dead coral areas was almost always able to prevent macroalgae from colonizing. Many species of smaller herbivores were able to escape the pressures of overfishing and effectively graze coral reefs in place of large keystone herbivores. This prevented macroalgae from aggressively colonizing places where live coral cover was very low. Initial diversity in coral reef fish species is important in degraded coral reefs to overcome anthropogenic pressures and stifle macroalgae growth. Further research must be done to determine whether grazing by small herbivores can shift a coral degraded area into a coral dominated area and how this will impact future coral resilience.

How Mangrove and Seagrass Habitats Can Save Corals from Climate Change

Today's oceans are heating and becoming more acidic due to the large amount of atmospheric CO_2 being absorbed by them. Rising ocean temperatures and ocean acidification are causing coral reefs to deteriorate. If oceans become too acidic, the hard carbonate foundation of corals could collapse, causing an immediate decrease in biodiversity and productivity in marine coral ecosystems. Many scientists are beginning to study corals living under extreme conditions such as near CO_2 vents and in areas with elevated temperatures or with a low pH. Mangrove and seagrass habitats have a wide range of temperature variation and light availability, but are still able to host large coral populations. Understanding coral reefs that have been able to tolerate unfavorable conditions is how scientists can learn how coral reefs will react under ocean acidification and rising sea temperatures.

Seagrass habitats can buffer corals from decreasing pH levels because using photosynthesis during the day and respiration at night creates a local carbonate chemistry that raises the mean pH of the surrounding water. Mangrove habitats also provide favorable conditions that may protect coral reefs against climate change. Camp *et al.* (2016) study how seagrass beds and mangroves provide important ecological services to protect coral reefs against climate change.

Seagrass habitats experienced higher pH and lower CO_2 levels than the outer-reef tested as the control. Calcification of corals occurred in the seagrass beds due to the ability of seagrass habitats to raise pH compared to outer-reefs. Mangrove habitats, however, experienced lower levels of pH, higher levels of CO_2, and lower levels of photosynthesis and coral calcification. Instead of buffering corals from ocean acidification, mangroves condition corals to withstand future acidic ocean conditions and naturally select corals that can withstand low levels of pH. Corals living in mangrove habitats are important to study because they can help scientists understand how corals will withstand future ocean acidification.

Corals' Final Refuge from Rising Ocean Temperatures in Moderately Turbid Waters

Increasing sea surface temperatures have lead to coral mortality in many tropical and subtropical areas. Both corals and the zooplankton that make up coral reefs are susceptible to stress due to rising sea surface temperatures, and periods of high seasonal temperatures can lead to coral bleaching, which often causes complete coral mortality. Since oceans are becoming less hospitable for corals, scientists are searching for refuges within oceans where coral reefs can survive and adapt. Previous studies have shown that coral reefs in areas close to shore with high amounts of turbidity have experienced less bleaching than have coral reefs in outer areas. Turbidity consists of small particles suspended in the water, which shades coral from high sea surface temperatures and shades corals from large amounts of sunlight. Its effectiveness depends on the size of sediments and the movement of water throughout the water column. Cacciapaglia & Van Woesik (2016) set out to discover turbid areas in the Pacific and Indian Oceans that could protect coral reefs from rising ocean temperatures.

Turbidity was predicted to protect 9% of coral reefs previously considered inhospitable for corals under rising ocean temperatures. The most widely distributed coral species in the Pacific and Indian Oceans, *Porites lobata,* will experience a 3% increase in reef habitat due to the effects of turbidity. The coral species *Dipsastraea speciosa* and *Montipora aequituberculata* will benefit the most from moderate turbidity within the Pacific and Indian Oceans due to the shading effects moderate turbidity has on corals. The amount of chlorophyll concentration in coral reefs also increased in areas with moderate turbidity. Turbidity refuges were found along the eastern and western coasts of Australia, New Caledonia, Central Indonesia, Eastern Vietnam, Northern Philippines, Japan, the Mozambique Channel, the Southern Red Sea, and The Persian Gulf. Generally, these turbid refuges were found between 20–30°N and 1–25°S. The researchers also found, though, that high turbidity could hinder coral reef growth. High turbidity can reduce light, constrict photic zones, and reduce rates of calcification. This scenario will most likely occur in Southern Malaysia and Indonesia. Overall, moderate turbidity can relieve pressure from coral reef habitats as ocean temperatures rise by providing shade and lowering ocean temperatures in near shore environments.

Conclusions

Coral reefs are facing many challenges as ocean temperatures rise and oceans become more acidic. Higher concentrations of CO_2 are stored in the ocean due to human activity. Corals are rapidly disappearing, causing other marine life to disappear too. Some refuges and management methods have been found to save coral reefs from these inhospitable environments. Moderately turbid waters and mangrove and seagrass habitats have been shown to protect some corals from massive bleaching events. Unfortunately, massive amounts of coral reefs are still dying off every day, and most of the damage is permanent. Scientists are struggling to save the coral reefs that are left. Only time will tell if coral reefs around the world can be saved and protected, or if it is too late to reverse the effects of ocean acidification and rising ocean temperatures.

References Cited

Bahr, K.D., Jokiel, P.L., Rodgers, K.S., 2016. Relative Sensitivity of Five Hawaiian Coral Species To High Temperature Under High-pCO2 Conditions. Coral Reefs. 35, 725–738.

Cacciapaglia, C., Van Woesik, R., 2016. Climate-Change Refugia: Shading Coral Reefs by Turbidity. Global Change Biology 22, 1145–1154.

Camp, E.F., Suggett, D.J., Gendron, G., Jompa, J., Manfrino, C., Smith, D.J., 2016. Mangrove and Seagrass Beds Provide Different Biogeochemical Services for Corals Threatened by Climate Change. Frontiers in Marine Science 3, 1–16.

Cheal, A.J., Macneil, M.A., Emslie, M.J., Sweatman, H., 2017. The Threat to Coral Reefs From More Intense Cyclones Under Climate Change. Global Change Biology 10.1111.

Gil, M.A., Goldenberg, S.U., Bach, A.L., Mills, SC., Claudet, J., 2016. Interactive Effects of Three Pervasive Marine Stressors in a Post-Disturbance Coral Reef. Springer 35, 1281–1293.

Kuempel, C.D., Altieri, A.H., 2017. The Emergent Role of Small-Bodied Herbivores in Pre-empting Phase Shifts on Degraded Coral Reefs. Scientific Reports 7, 10:1038.

Smith, J.N., De'ath, G., Richter, C., Cornils, A., Hall-Spencer, J.M., Fabricius, K.E., 2016. Ocean Acidification Reduces Demersal Zooplankton that Reside in Tropical Coral Reefs. Nature Climate Change 6, 1124–1130.

Wen, C.K., Bonin, M.C., Harrison, H.B., Williamson, H.B., Jones, G.P., 2016. Dietary Shift in Juvenile Coral Trout (*Plectropomus maculatus*) Following Coral Reef Degradation From a Flood Plume Disturbance. Coral Reefs 35, 451–455.

Climate Change and Endangered Species: The Global Effects

Kelsey D'Ewart

Climate change is becoming increasingly problematic and its effects are more apparent every year. It is commonly discussed in the context of agriculture, rising sea level, melting polar ice caps, and global temperature increase. However, it is less often talked about in reference to its detrimental effects on endangered species. Polar species such as polar bears are often mentioned in conjunction with ice melting, but the endangerment of other animals in different habitats and locations is often over looked. Climate change can have impacts on species habitats, distribution, phenology, physiology, and more. More importantly, the effects of climate change on one species have the potential to easily create a cascade effect on the entire ecosystem, putting entire regions at risk. The longer the global effects of climate change are ignored the more it will become exponentially worse in the future.

This chapter provides an overview of many different species and the issues they have encountered as a result of climate change. Articles are summarized that include many different types of habitats. The discussion of a diverse range of habitats is important in order to obtain a broad view of the effects that climate change is causing, rather than simply looking at one region, such as just the polar region. The different habitats include polar regions, mountain rivers, tropical islands, wetlands, temperate forests, hot deserts, temperate deserts, and temperate shrubland, The articles cover habitats in North and South America, Africa, Asia, and Europe. The types of animals include large and small mammals, reptiles, and birds. This wide array of climates and locations is discussed in order to gain a more in depth understanding of how many different ways climate change can be detrimental to different species.

One method of studying these changes discussed in several of the articles is species distribution models (SDMs). SDMs, which have been critical in conservation biology over the past several years, help researchers model predicted species population sizes as well as distribution, characteristically for as much as 100 years into the future. They do so by using previously collected data along with current data and then extrapolate future data using these. SDMs have become increasingly popular and are useful tools in predicting the effects of climate change as well as aiding in attempts to implement conservation efforts for endangered species. SDMs are used throughout many of the articles in this chapter and give a promising option for researchers to continue making strides to combat the effects of climate change on endangered species.

Climate Change and Land Use Effects on the Red-Cockaded Woodpecker

Climate change has been a pervasive issue when looking at the health and protection of endangered species. However, land-use has also been a significant factor in the decreasing population size of endangered species. Together, climate and land-use change affect habitat, behavioral patterns, phenology, and many other parts of many species' lives. This is especially relevant to species that have specific habitats, dietary needs, or both. If these issues are not addressed, the risks of endangered species becoming extinct drastically increase. Bancroft *et al.* (2016) studied the affects of land use and climate change on the Red-cockaded Woodpecker (RCW) using a modeling system that allows them to analyze different predicted land-use and climate change scenarios up until the year 2100. This study looks at the RCW population in Fort Benning, which includes specific pine forests vital to the RCW survival. The authors looked at three different potential future conditions– conservation, convenience, and worst-case– to determine what types of major changes might occur in the RCW population over the remainder of the century.

Bancroft *et al.* used RandomForest package in the statistical package R, to build the predicted habitat models. They utilized the potential training maps at Fort Benning to predict how land use might change in the RCW habitat. In each of the three scenarios (conservation, convenience, and worst-case), several areas were further developed in the model at each ten-year point. In the maps for each scenario, new developments were added so that by 2100 all of the predicted development existed on the same map. In order to predict climate change, Bancroft *et al.* first used the ecosystem model LPJ-GUESS to assess the potential changes in vegetation throughout the habitat using multiple scenarios. Additionally, they explored the different changes in rainfall that could occur due to climate change. Lastly, they made a population simulation using HexSim that took into account both the projected climate and land use change. In the model, 800 RCW's were implemented randomly throughout the habitat.

Bancroft *et al.* determined that the specific land-use change scenario (conservation, convenience, and worst-case) was a significant factor on how the RCW population changed. The conservation population remained similar over the 100 year time period, while the convenience scenario experienced a slight decline in the RCW population, and the worst-case scenario population was greatly impacted, showing a large decline. Overall there was a significant difference in population size between all three scenarios at the end of the simulation. However, climate change was found to have minimal effect on the populations.

Although climate change can be extremely detrimental to endangered species, Bancroft *et al.* remind us that land use change is another important factor to take into consideration when discussing conservation efforts for individual species. This is especially vital when looking at species with highly specific habitats, such as the RCW. It is not impossible to prevent these species from going extinct, because land use change can be done strategically and minimally so as not to greatly affect the species living in these habitats.

Arctic Climate Change's Effect on Caribou Migration

The freezing and thawing patterns in the Arctic have been increasingly affected by increasing global temperatures, resulting in later freezing and earlier thawing. This is forcing phenology changes in many Arctic species, particularly in response to a decrease in frozen bodies of water, which can lead to longer, more strenuous, and more dangerous migrations that can result in higher mortality rates. Leblond *et al.* (2016) tracked the ice thawing and freezing times for bodies of water in the migration path of caribou, *Rangifer tarandus*, in Northern Quebec from 2007 to 2014, allowing them to determine if the change in ice melt was affecting the caribou's migration route. Their hypothesis was that the caribou would travel extra distances in order to avoid swimming in water that was not completely frozen. They assessed four different parts of the migration: previous data for freezing trends, the caribou's response to the change in freezing trends, fine-scale caribou behavior and phenology, and possible future movement using climate change projections.

To study the movements of the caribou herd Leblond *et al.* used GPS collars on 96 caribou, the majority of which were females because the migration being observed was to breeding grounds that males do not frequent. The bodies of water assessed were chosen using the route of the caribou's migration to determine the lakes, rivers, and reservoirs most encountered. Thawing and freezing patterns were determined using moderate-resolution imaging spectroadiometer (MODIS), which was also used to estimate the proportion of ice to water on bodies of water the caribous would typically cross. Thawing and freezing trends were then extrapolated using linear regression. GPS and turning angle (a metric to determine if the state of bodies of water altered the migration route) were used to study the caribou behavioral patterns. Additionally, future ice freezing and thawing patterns were projected up for the year 2070 to predict how the caribou's phenology may change.

Leblond *et al.* found that from 2000 to 2014 there was no significant change in the thawing and freezing dates of the bodies of water. They found that if there was absolutely no ice available for caribou to cross, they would swim distances as long as 25 kilometers. The number of water crossings remained constant regardless of whether it was a late or early freezing year. However, when ice was not available the caribou would "pause" to either rest or determine if it was safe to cross, which took both extra time and energy. The caribou were able to move more efficiently on ice, and had fewer turning angles than in open water. Future projections showed that only a small percent of spring migrations would not occur if water had thawed 10–15 days sooner. However, projections in 2070 showed that up to 36% of crossings might be greatly hindered if the earlier thawing trend continues.

Although migration patterns were not significantly altered in the small time period of this study, the future projections show the possibility of a much smaller frozen period. This will lead to more caribou swimming or detouring, which could lead to much larger energy expenditure as well as potential drowning. For the moment a slight shift in migration time will be enough to help the caribou, but in the future the global temperature increase and ice thawing could become a serious problem for the caribou and many other species.

Warming Air and Sand Temperatures Affect Sea Turtle Hatchling Sex Ratio

Increasing global air temperatures are also increasing temperatures of soil, water, and sand, all of which influence breeding patterns and development of marine ectotherms such as sea turtles, which rely heavily on outside external heat sources to regulate their own body temperature. This is even more pronounced in species where sex is determined by temperature. Laloe *et al.* (2015) looked specifically at several endangered sea turtle species' breeding patterns and hatchling sex ratios, and found that male hatchlings are becoming scarcer, putting the species at risk.

Laloe *et al.* studied St Eustatius Island in the Caribbean Sea, a prominent breeding ground for green hawksbill, and leatherback turtles. They utilized previous rainfall data as well as air and sea temperature to look at past changes. Additionally, they acquired air temperature projections from United Nations Development Programme. Tinytag Plus 2 Loggers were used to track sand temperatures in the species' various nesting beaches. Both temperature and sand albedo– a measure of the amount of light reflected– were measured. Incubation temperature and sex ratio data from 2014 were used to calculate potential sex ratios, and 29°C was used as the "illustrative" temperature for the sea turtles.

Laloe *et al.* found that the greens and hawkbills tended to breed when the sand temperature was at its highest. They also found that rainfall and spring tides could have an effect on sand temperature for the season. The depth at which the sea turtles laid their eggs was measured because and temperature varies at depth. There was no significant effect of sand albedo found, meaning that light reflected was not a factor in the temperature of the sand. However, Laloe *et al.* found a significant relationship between air and sand temperature. They then modeled both past and future air to sand temperature relationships, and found that sand temperatures will most likely continue to increase. They extrapolated that female sex ratios will rise to above 95% in the coming years for the majority of the turtle species.

As the ratio of females to males continues to rise, the possibility of extinction for many of these turtle species increases. Some nesting grounds have even shown up to 100% female hatchlings. Although in theory a larger female to male ratio may be beneficial for the present (more females leads to an increase in number of hatchlings) this could ultimately be unsustainable. Natural occurrences such as an increase in rainfall could lead to a higher male hatchling rate, however rainfall can also damage nests and cool sand temperatures to a fatal level. Despite all this, Laloe *et al.* found that there is a possibility for the turtle species to adapt; such as changing their mating seasons forward or backward depending on the species.

Using Blunt-Nosed Leopard Lizards to Predict Future Drought Effects

Droughts have become more common and severe as climate change has persisted, leading to a drastic change in habitats, causing many issues for the animals living in dry, hot climates. However, Westphal *et al.* (2016) suggests that although these more persistent droughts are not ideal, they give scientists and opportunity to model potential future effects of more severe droughts. They used the blunt-nosed leopard lizard to study future drought effects by counting how many appeared in various surveys during a persistent drought in the Southwest United States. They then used these data to extrapolate the future consequences of lengthy droughts, and to determine the relationship between blunt-nosed leopard lizard juvenile recruitment and winter precipitation.

Westphal *et al.* chose a xeric climate as they are the most susceptible to droughts, and because they have highly specific habitats that individual species rely on. They used juvenile recruitment as a measure for species persistence, and used the drought indices as well as the Normalized Difference Vegetation Index (NDVI) in their short-term study to create a future model. They did a presence/absence style survey, and did a survey for only one season.

The results indicated that there is a significant relationship between increased winter precipitation and number of juvenile Blunt-Nosed Leopard Lizards spotted. Additionally, the NDVI index was higher in areas where lizards were found. Some sites surveyed that historically housed significant lizard populations had no lizards during the survey. This absence of lizards indicates a significant future problem in lack of breeding, potentially caused by severe droughts. Given these results Westphal *et al.* concluded that droughts as a result of climate change will be a significant risk factor for many species in the future. Although other studies must be done to better predict drought effects there is already ample evidence that xeric habitats and their species are at great risk.

Kelsey D'Ewart

Predictive Models for Aquatic Mammals Using Pyrenean Desman

Many researchers have been turning to species distribution models (SDMs) rooted in the ecological niche theory to help predict how climate change will affect endangered species in the near future. SDMs are helpful in determining narrowing species ranges, new environmental factors, and other changes due to climate change and development/land use. Although SDMs are not always accurate they are one of the best types of models available to estimate future shifts, but they have seldom been applied to semi-aquatic mammals. As a result, fewer conservation measures have been implemented for these species. Additionally, hydrology is rarely considered in freshwater ecosystem model of mammals even though it is an important factor. Charbonnnel *et al.* (2016) used a Soil and Water Assessment Tool (SWAT) in combination with a SDM and previous surveys to predict the future distribution and range shift of the Pyrenean desman; a small, threatened, semi-aquatic mammal located near rivers in the Pyrenees Mountains. They studied whether or not hydrological variables help predict distribution, whether or not the Pyrenean desman distribution has shifted over time, and whether or not SDMs are accurate predictors.

Charbonnel *et al.* used an integrative model approach, including SDM and SWAT, by incorporating whole stream network data on a large spatial scale to assess the changing range of the Pyrenean desman. They studied 1222 km sampling sites, which included 500 m of riverbed transects, and recorded the presence or absence of the species. SWAT was used to determine monthly stream flow in the entire French Pyrenees stream network. Various other climate variables including daily rainfall, maximum and minimum air temperature, solar radiation, humidity, and more were also taken into account. Non-natural factors such as human disturbance also went into the model. Data were used from 1990 to present day.

In historical surveys the species was detected at 81% of the sampling sites, however current sampling showed it at only 46% of the sites, and it was detected at 29% of previously inhabited sites.

The authors also found an increase in air temperature, particularly at the higher elevations, as well as a decrease in annual rainfall, which affected population distribution. Additionally, stream flow was, on average, lower in the current period compared to historical data. Land use and human disturbance was found to have an insignificant influence. The model showed a decrease in habitat by as much as 20% in the past 20 years, and predicts a loss of up to 20% more in the near future. However, the model gave an inaccurate estimate for the current range, indicating that it may not be the best future predictor. Charbonnel *et al.* believed that it would over predict habitat suitability in the future.

Overall the Pyrenean desman habitat has drastically decreased in the past 25 years, and that it will most likely decrease at an even faster rate than the given SDM model presents. The authors also noted that their model indicates that other factors not listed may have a significant affect on the habitat range, as factors such as biotic interactions are often not considered in SDMs, even though they can often have a larger affect than climate change.

SDMs Can Help Develop Conservation Methods for Oriental White Storks

Increasing global temperatures are becoming more problematic because many species are not biologically able to adapt past a certain threshold. In conjunction with rising temperatures, human population expansion, and increased human land-use has shrunk habitats and isolated many species as their ideal habitats shrink around them. Species Distribution Models (SDMs) are increasingly prevalent in the study of endangered species, and are useful when creating potential conservation efforts, as they can account for many variables including both climate change and land-use change which are often two main factors in further endangering species. Additionally, SDMs can be applied to every type of habitat given the correct set of parameters. Recently, Zheng *et al.* (2016) created an SDM for the rare oriental white stork species, which is found in wetlands in Northeaster China.

The oriental white stork is a rapidly declining flagship species, suspected to have only around 2500 individuals left: it dwells in wetlands that are rapidly being deforested and made into cropland. However, recent efforts made by the Chinese government have helped partially restore the population, emphasizing that if conservation efforts are implemented this species could eventually be removed from the endangered species list. Zheng *et al.* used MaxEnt, an SDM program, and historical data, to predict three different conservation models to determine which were most effective. The authors studied existing data from the Sanjiang Plain region of stork presence, breeding rate, and present and future environmental factors. Three conversation scenarios were developed. First, reclaiming cropland back to wetland (RCTW) converted current croplands back into suitable habitats for the storks, as this is the stork's first choice habitat. Second, establishment of nature reserves (EONR), which involved the government partitioning wetlands, meant for the storks and other animals in that ecosystem. Third, installing artificial bird nests (ABN), which included relocating storks into a habitat with manmade nests. The results indicated that each climate variable tested had a potential significant impact on the stork habitat. Predicted moderately suitable habitats were found to decrease by 2.71% by 2050. Given the RCTW scenario, moderately suitable habitat only decreased by 2.55%. Under EONR, moderately suitable habitat increased by 0.16%. There was no increase found in moderately suitable habitats for ABN.

The MaxEnt program showed that climate change coupled with no conservation efforts would negatively affect the storks in the future, with temperature and early breeding precipitation being important factors. Conservation strategies have the potential to save both feeding and nesting grounds. RCTW was positive for the storks but at a large human cost for food production. EONR allowed for a more natural habitat, which was said to be able to provide a "unique opportunity" for the storks. ABN was not considered a useful conservation option given the results of the study. The SDM models allowed for two different conservation strategies to be tested, and now must be accepted and implemented by local governments if they are to do any good.

Reptile Distribution and Vulnerability due to Climate Change in Tanzania

Although the literature on endangered species and climate change is vast, few studies explain the relationship between climate change and reptiles. This is problematic because reptiles rely heavily on ambient temperature for physiological regulation. For this reason, Meng *et al.* (2016) looked at the reptile diversity in Tanzania, a country that has many different habitats and landscapes. Tanzania consists of habitats including savannahs, alpine grasslands, mangroves, and more, so it allows for studying many different species all in relatively close proximity. Meng *et al.* focused largely on the Eastern Arc Mountains because they historically act as refuges, as well as a habitat where speciation has occurred during climate change. Their goal was to more clearly map Tanzanian reptile distribution and the impending threats due to climate change, as well as demonstrating to policymakers that the global IUCN Red List methodology for endangered species may have to be altered and prioritized differently.

Meng *et al.* collected data from two sources, the IUCN Red Listing Workbook, and IUCN Red List assessments. Range maps were used from 188 amphibians, 356 mammals, and 1046 bird species in order study the spatial congruence in reptiles. In order to determine if a species had high sensitivity to climate change Meng *et al.* used several factors including specialized habitat/microhabitat requirements, species tolerances that might be surpassed by climate change, environmental triggers that may accelerate climate change, inter-species interactions, and more. Then, change was projected in four areas: mean temperature change, total precipitation change, change in temperature variability, and change of precipitation variability. Then sensitivity, adaptability, and exposure to climate change were put into the model to determine which reptile species were particularly vulnerable to climate change. Results indicated that reptile species diversity was closely linked to amphibian richness. Thirteen percent of the reptiles studied were threatened or endangered, most of which are found in the Eastern Arc Mountains. The most impactful threats not related to climate were land-use and agriculture; the latter has been slowly expanding into the Eastern Arc Mountains. The model projected that by the year 2070, 68% of species will be at high-risk from to climate change.

The study found that agriculture was even more threatening to reptile species than originally thought, as the Eastern Arc region studied has suffered more than 75% deforestation. Animal export and trade is also becoming an increasing threat, and Tanzania is a major chameleon-exporting country. Meng *et al.* also determined that the IUCN Red List must not consider spatial proprieties for non-climate-threatened species with climate-threatened species, as they do not mirror each other. The results indicate that new policies and better management must be put in place if the endangered reptile species are going to be conserved long term. Meng *et al.* was able to help narrow the knowledge-gap of reptile biodiversity and species richness, allowing the Tanzanian government to have more tools as their disposal to improve conservation efforts.

Climate Change Vulnerability Assessments of United States Sagebrush Habitat

One of the shortcomings of the effects of research of climate change on endangered species is that most studies focus on a single species climate envelope. Many new studies use climate change vulnerability assessments (CCVAs), which incorporate more than one climate envelope to predict future effects on species and to help conserve landscapes. *Balzotti et al.* (2016) incorporated multiple species and their respective climate envelopes as well as other environmental stressors to project risk maps for mesic (Strawberry) and xeric (Sheeprock) sage-grouse habitats in Utah. They chose these species because sagebrush habitats have been reduced throughout these sage-grouse ranges, more so than most other ecosystems in the United States. Additionally, sage-grouse are a good species for risk assessment because they have a large distribution, recent decline in suitable habitat, very specific habitat needs, and many prior studies to gain information from. Additionally, measures put into place for the conservation of sagebrush habitat benefit many species, not just the sage-grouse.

Balzotti *et al.* utilized two spatial scales to determine sage-grouse climate vulnerability, incorporating 23 bioclimatic variables. These included mean annual temperature and precipitation, and encompassed a wide range of climate effects. Invasive conifers and invasive annual grass were taken into account. Habitat variables such as fire history, invasive species encroachment, and human population growth were also implemented in the model. These non-climatic factors were considered because if the climate envelope for sagebrush suffered a change then these factors would be even more impactful.

Balzotti *et al.* determined that sage-grouse and similar species will be greatly affected by climate change in the near future. Their work indicated that one of the most prominent risks was droughts, specifically for habitats at higher elevations. Sheeprock sage-grouse were determined to be at higher risk than Strawberry sage-grouse due to a larger loss in their preferred habitat, as well as conifer encroachment. Additionally, Strawberry sage-grouse were found to be more connected to other sage-grouse habitat which may lead to more genetic diversity and adaptation. Balzotti *et al.* aimed this study towards Federal land management agencies so they would focus on more important, long-term conservation efforts for landscapes and species that can more easily be maintained.

Conclusions

Although many different species and habitats were explored throughout this chapter there are several clear trends that cannot be ignored. First, climate change is a prevailing threat affecting many endangered species, and if measures are not made towards reducing these risks then many species have a large chance of going extinct. It is also clear that climate change cannot be the only factor considered when taking action in protecting endangered species. Human land-use is becoming increasingly destructive to many habitats that are already quickly diminishing because of climate change, and it is necessary to consider climate change in conjunction with land-use in order to truly make progress in protecting endangered species. These themes hold true regardless of the species, climate, or region. From the Pyrenees Mountain range to the tropical islands of the Caribbean these are both important factors in preventing the loss of already diminishing diversity.

References Cited

Balzotti, C., Kitchen, S., McCarthy, C. Beyond the single species climate envelope: a multifaceted approach to mapping climate change vulnerability. Ecosphere 7, 1–23.

Bancroft, A. B, Lawler, J. J., & Schumaker, N. H., 2016. Weighing the relative potential impacts of climate change and land-use change on an endangered bird. Ecology and Evolution 6, 4468–4477.

Charbonnnel, A., Laffaille, P., Biffi, M., Blanc, F., Maire, A., Nemoz, M., Sanchez-Perez, J.M., Sauvage, S., Buisson, L., 2016. Can Recent Global Changes Explain the Dramatic Range Contraction of an Endangered Semi-Aquatic Mammal Species in the French Pyrenees? PLoS ONE 11. 1–21.

Meng, H., Carr, J., Beraducci, J., Bowles, P., Branch, W., Capitani, C., Chenga, J., Cox, N., Howell, K., Malonza, P., Marchant, R., Mbiliny, B., Mukama, K., Msuya, C., Platts, P., Safari, I., Spawls, S., Shennan-Farpon, Y., Wagner, P., Burgess, N. Tanzania's reptile biodiversity: Distribution, threats and climate change vulnerability. Biological Conservation 204, 72–82.

Laloe, J.O., Esteban, N., Berkel, J., Hays, G. C., 2015. Sand temperatures for nesting sea turtles in the Caribbean: Implications for hatchling sex ratios in the face of climate change. Experimental Marine Biology and Ecology. 92-99.

Leblond, M., St-Laurent, M., Cote, S. 2016. Caribou, water, and ice-fine-scale movements of a migratory arctic ungulate in the context of climate change. Movement Ecology 4:14.

Westphal, M.F., Steward, J., Tennant, E.N., Butterfield, S.H., Sinervo, B. Contemporary Drought and Future Effects of Climate Change on the Endangered Blunt-Nosed Leopard Lizard, *Gabmbelia sila*. Plos One. 1–9.

Zheng, H., Shen, G., Shang, L., Lv, X., Qiang, W., McLaughlin, N., He, X. Efficacy of conservation strategies for endangered oriental white storks (*Ciconia boyciana*) under climate change in Northeast China. Biological Conservation 204 367–377.

Section III—Energy

Shale Oil Fracturing and Climate Change

Catherine Parsekian

The process of using shale oil in place of coal or conventional crude oil began around 1950s when World War II caused a shortage of gasoline. By the 1980s, shale oil production hit its peak, particularly in China, the United States, and Russia. However, many major issues have arisen with the use of this alternative fuel source, including environmental and health impacts. Many scientists argue that there is a lack of regulation surrounding the entire fracking process, an issue that could have extremely detrimental environmental effects if it is not properly addressed. The process of fracturing is very risky, and if it is not performed correctly; inadequate wastewater disposal, methane gas leakage, and irreversible geographical impacts can arise quickly. Many studies show that shale oil is a good alternative to coal or crude oil, but only if the fracking is performed properly.

There is much unknown about the impacts of fracking for shale oil on people and the environment, both locally and globally. For example, some studies have shown that people living is fracking communities experience damage to both their mental and physical health. These studies, while extremely limited, suggest a need to further study the effects that shale oil production could be having on local communities. There is also a need to regulate the way in which companies are able to pick which areas to frack, especially in areas where a property owner can turn away a fracking company only if they own the mineral rights to their land. Many studies are also showing that companies have a poor ability to measure an area's potential shale quality, forcing them to waste time and resources with no guarantee that they will find useful shale energy. Finally, there is an overwhelming fear that shale production could not realistically be an alternative energy to crude oil because there is not enough of it available to support the demand.

Environmental Issues Related to Fracking and their Potential Solutions

Meegoda *et al.* (2016) researched various environmental issues related to the process of shale fracturing using data from the Marcellus shale formation. These major impacts include methane gas leaks, the triggering of earthquakes, and wastewater disposal. Meegoda *et al.* argued that these environmental impacts caused by fracking need to be addressed and properly regulated if shale oil is to be used as an alternative source of relatively clean energy. Without policing these problems and fixing them, fracking will simply result in new detrimental impacts to the environment, making it no better than coal or traditional natural oil production. In order to deal with the issue of methane gas leakage, Meegoda *et al.* emphasized the need to properly train pipe and well builders. The authors point out that the integrity of the pipes and wells is extremely important because any cracks or bad cement can lead to emissions of methane. In addition, the installation of methane monitors and alarms would be very useful.

Earthquakes pose the least likely of threats, but also the most dangerous in the rare case that they occur. While there is some evidence that more earthquakes occur in fracking areas due to well drilling, they are not large enough in the Marcellus shale so far to be felt by humans. One preventative measure that the authors recommend is the prohibition of underground injection of waste fluid, as well as properly vetting potential fracking areas for geographical fault lines. Finally, the authors argue that any fracking fluid produced needs to be properly treated and extracted from the areas so as to minimize the impact of waste disposal. Regulations need to be introduced to the shale oil industry so that wastewater is properly treated or disposed of. The best way to do this would be reusing the fracking fluid and treating any waste in a private treatment plant. The authors also introduce many chemicals that could be used for better water treatment and disposal, including acetaldehyde, ethylene glycol, and ammonium persulfate. If these changes and regulations were to be introduced to the shale oil industry, it could potentially be a much cleaner, safer alternative to coal.

Overview of Water Use and Management in Shale Oil Basin of North Dakota

Horner *et al.* (2016) recently put together a well-formed summary of the water use and management associated with communities surrounding shale oil basins, particularly the Bakken deposit in North Dakota. Through their extensive comparisons of various data collections, including the North Dakota Department of Mineral Resources, the US Geological Survey, and the North Dakota Industrial Commission, they were able to compare the water usage changes from 2005 to 2012. The authors found that the amount of water annually consumed for hydraulic fracking was five times more in 2012 than it was in 2008. Homer *et al.* also found that the annual rate of well installation has quadrupled. Another value they measured was water use intensity, and they found that roughly 0.16–0.33 gallons of water are used per gallon of crude oil produced in areas fracked once.

Horner also gathered information regarding domestic water use. As hydraulic fracking becomes lucrative and popular in an area, the number of workers per well also grows. The authors estimate that the amount of water consumed by workers on wells in the Bakken basin is at least 2.19 billion gallons. Furthermore, the use of fracking depletes water sources. Many developers use groundwater because it is closer and cheaper to transport. However, groundwater sources are not enough; in 2012, 4.3 billion gallons of water were needed for fracking in the Bakken basin, but only 3.7 billion gallons of groundwater were permitted for withdrawal. The next source of water is surface water, which the authors note is plentiful in North Dakota. Finally, they note that wastewater levels of the area have exceeded the volumes of the water actually injected into the wells for fracking. All of the information they gathered relates solely to the Bakken basin in North Dakota, and the authors note that each basin is extremely variable, making it difficult for them to generalize their findings. Overall their study concludes that for this particular basin, there appears to be a need to balance the level of water being used in relation to the amount to shale oil being produced. Horner suggests that not enough oil is being produced to justify the amount of water being used.

Health Issues Associated with Hydraulic Fracturing Communities

Sangaramoorthy and associates (2016) gathered qualitative evidence surrounding the potential mental and physical impacts that fracking is having in Doddridge, Maryland. Overall, their very limited study does point to many possible issues revolving around the use of fracking. While the concept of hydraulic fracking has been seen as an alternative source of energy developed to alleviate fears of a crude oil shortage, the process itself has many concerns. For example, the method for drilling into areas like the Marcellus Shale is very different than the old method of drilling for oil. It involves disrupting a much larger area of land and requires the use of a pressurized mixture of water, sand and chemicals injected into the well in order to release the shale oil. This process is new and there have been few quantitative studies done regarding the physical health impacts fracking has on its communities.

This study utilized observation and focus groups in order to gather the general social and mental health climates of fracking town Doddridge, Maryland. Sangaramoorthy *et al.* conducted two focus groups with a total of thirteen participants, which is extremely few and makes the overall study's findings potentially biased. However, the general worries and opinions expressed by participants still raise concerns regarding the treatment and general health of people who occupy fracking areas. With regard to mental health, numerous by-products of fracking were cited as stress inducers for the community. These included the rise in insurance costs for homes, the plummeting of home values, and insecurity about the future. In addition most people felt that the entire process of fracking was basically seizing their land without permission, since mineral rights and surface rights are separate. While many people personally interviewed did not claim any major physical health issues, almost everyone knew people who suffered from nosebleeds, sore throats, skin rashes, chemically induced asthma, and headaches. While these mental and physical health issues were not proven to be a wide-spread issue in fracking areas, this study shows that while fracking may be a viable alternative to traditional crude oil drilling, more studies need to be done to measure the health effects that shale oil production is having on citizens subjected to it.

Effect of Fracking on Traffic and the Resulting Environmental Impacts

According to a study done at Newcastle University by Goodman *et al.* (2016), while many environmental impacts of fracking have been studied, traffic and road effects have been largely ignored. The team developed the Traffic Impacts Model (TIM), a new system that estimates traffic and traffic-related impacts that fracking has on greenhouse gas emissions, local air quality, noise pollution, and pavement deterioration. TIM uses hypothetical scenarios to quantify these impacts, considering short-term effects on single areas, as well as extrapolating data to look at long-term effects on larger regions. This program attempted to consider all possible environmental impacts of the fracking process, including construction and drilling, the actual hydraulic fracturing, flowback treatment, and miscellaneous operations and activities of those working at fracking areas.

While this study was extensive and well thought-out, the overall results showed the environmental impacts of fracking related to increased traffic are far over-shadowed by the effects of the actual hydraulic fracturing process. While Goodman *et al.* successfully showed that the traffic impact of a single well can result in large increases in NO and CO_2 emissions, noise pollution, and road pavement corrosion, these results are only true for extremely short periods of time at peak oil production. In addition, Goodman *et al.* discussed that long-term adverse effects were negligible when compared to all traffic and industrial activities in general. Overall, the study concluded that traffic-related impacts of fracking are just adding to all the other industries whose traffic and production processes are having negative environmental impacts on the planet, including increased greenhouse gas emissions, noise pollution, air quality, and road pavement destruction. It seems clear that while there may be severe environmental implications of the actual hydraulic fracturing process, the impacts related to traffic are insignificant in comparison. While Goodman's study was impressive in its use of TIM and the ability to look at the environmental impacts of traffic in hypothetical situations, this study simply emphasized the need to find a new fuel source that does not have such a high environmental impact.

Light Tight Oil Production Causes Skewed Evaluations of US and Global Oil Production

According to James Murray of the University of Washington, most major agencies and prognosticators have underestimated the future demand for oil production. One of the main reasons, Murray argues, is a limited understanding of why there was an increase in global oil production in the last four years. Light tight oil (LTO) production is a very costly and expensive way of producing oil, the overall process costing about $6–$9 million per well (Murray 2016). Yet in 2011, world oil production surged, mainly due to a surge in LTO production. It was not because of new technology; LTO increased because the market price of oil was high enough to support the expensive process. However, as time has gone on, the price of oil has dropped significantly, causing LTO production to drop in conjunction with it.

As the price of oil drops, oil companies are no longer willing to spend extra money to search for new wells and basins. They are also less willing to spend time on expensive projects like fracking and shale oil production when it threatens their investors' interests. Not only are oil companies not willing to spend money on finding alternative fuels when the return on these projects are often extremely low, there are fewer viable alternatives to conventional oil than previously predicted. Murray argues that not only is there not enough oil to meet demand of the future, he suggests that there is not enough alternative fuel to meet the demand either. Even though the production of LTO surged, it quickly became apparent that it was both extremely expensive and low yield. Experts had estimated that LTO would be a great alternative way of producing oil, but that was shown to be an overestimate. Murray's overall argument takes issue with the current evaluations of oil production as well as expresses the need to rely on other forms of energy that do not include conventional oil, fracking, coal, and shale oil.

Proper Assessment of Shale Oil Quality

According to the results of a study done in China by Li *et al.* (2016), there is no method for measuring oil potential in shale reservoirs that includes both residual oil contents in the rocks as well as hydrocarbon expulsion and migration conditions. Li and his colleagues developed an index for determining oil potential. If the index is greater than zero, then some of the oil has migrated to external reservoirs, which means that it has poor shale oil potential. Li *et al.* argue that because current methods include absorbed, as well as free hydrocarbons, they are overvaluing the shale oil and not looking at oil that can readily be used. The method developed in this paper has multiple parameters and is a more comprehensive measurement since it takes into account oil saturation, free oil content, and shale oil expulsion.

Working in Biyang depression, located in the Nanxiang Basin in eastern China, these scientists performed analyses on the shale rock, using a combustion oven, measuring carbon dioxide production to calculate total organic carbon (TOC) in terms of milligrams of hydrocarbon per gram of rock; the results "represent the difference between total oil generation volume and pyrolysis free-hydrocarbon". The important concept that this paper brings to light is that while shale might have a very high total organic carbon count, it could also have an equally high absorption capacity for oil, and also overall poor shale potential. Studies such as this to make measurements more accurate are helpful in determining if shale oil can be used as a reliable and valid alternative to petroleum.

Fracking Fluid Comparisons

A study done in Colorado by Kim *et al.* (2016) attempted to compare two fracking fluids to see if changing their formulas would alter how the produced water would need to be treated for reuse. The experimenters studied samples sporadically gathered over a 200-day period from two wells located in Northeast Colorado. The process of fracking basically means injecting frac fluid at high pressure into the ground in order to forcefully extract oil or gas. This study focused on comparing two possible fluids to be used during fracking; fluid A comprising a residue-free polysaccharide, and fluid B made of derivatized guar. Not only were Kim *et al.* interested in these two fracking liquids, they were also interested in well age and how these two variables could effect produced water quality. As measures of water quality they looked at pH, concentrations of zirconium, aluminum, and potassium, and other chemicals.

Kim *et al.* showed a significant difference between fracking fluids in compound concentration as well as water pH. Fluid A had a mean pH of 7.22, while fluid B had one of 7.13. While these values seem very close, Kim *et al.* argue that because the values were significantly different, there is a strong influence of the formation on buffering and determining produced water ionic composition. They also saw a significant positive correlation between the key ions and TDS. The authors also measured temporal variability, or any difference in water quality that might have depended on the time the water was collected from a wall. They argue that barium, boron, chloride, potassium, sodium, strontium, sulfate, and silicon all indicate that the two fluids had different initial concentrations for produced water quality. Overall, it appeared that their second fluid appeared to return higher quality water, meaning that they would need less effort to be treated for reuse. This study touches on the importance of considering the repercussions that come from fracking, including the flowback of water, and the need to reuse that water in order to make fracking a feasible and efficient way of finding untapped oil.

Future Models Used to Predict Fracking Impacts on Climate Change

A study done by Sharma *et al.* (2016) utilized various watershed models in order to predict how current fracking trends could affect future stream low flow, the flow of water in a stream during dry weather. The process of hydraulic fracking includes the withdrawal of water from streams and rivers, which could lead to ecological stress adversely affecting aquatic life and municipal water supplies. The overall purpose of this study was to use various models, including the Soil and Water Assessment Tool, the Coupled Model Intercomparison Project Phase 5, Max Planck Institute Earth System Model, and three Representative Concentration Pathways to evaluate current fracking trends and attempt to predict future impacts of these trends on the climate. The authors gathered all of their data and information from the Muskingum watershed, which is located in eastern Ohio and is one of the largest watersheds in the United States.

Sharma *et al.*'s experiment was very thorough and extensive, yet its findings were inherently inconclusive and speculative. The authors determined that there would be negligible impact of hydraulic fracturing on mean monthly and annual stream flow. Furthermore, their findings indicated that future climate change might not have additional adverse impacts if hydraulic fracking trends remain the same. It was also suggested that the time between 2021 and 2050 will be a critical time period for sustainable water management, and their models suggest that planners need to devise policies that include preservation of water resources in light of future climate change scenarios. Most of their conclusions were limited due to the variability of stream flow and the overall uncertain nature of future climate conditions and scenarios. While the study is inherently hypothetical, it did a good job exploring a variety of scenarios for the region and utilizing many kinds of model simulators to determine the most likely effects and outcomes of fracking on water levels when potential climate change was taken into account.

Conclusions

Shale oil has given many people hope that there is a viable alternative to coal and crude oil to meet the transportation needs of the world. However, many studies show that even if companies were able to harvest shale at an efficient and safe level, there may still not be enough to meet demand. Even if there were enough shale oil available, the cost of producing it would make it far too expensive for realistic commercial use. Studies do show that if the process of fracking is done in a safe and careful way, it has the potential to be a relatively cleaner source of energy than coal and gasoline. While this knowledge may be encouraging, more studies need to be done to prove that the use of fracturing will not just cause other adverse results such as water resource depletion or geological land ravaging.

References Cited

Goodman, Paul S., Galatioto, Fabio, Thorpe, Neil, Namdeo, Anil K., Davies, Richard J., Bird, Roger N. Investigating the traffic-related environmental impacts of hydraulic-fracturing (fracking) operations. Environmental International 89-90, 248–260.

Horner, R. M., Harto, C. B., Jackson, R. B., Lowry, E. R., Brandt, A. R., Yeskoo, T. W., Murphy, D. J., and Clark, C. E., 2016. Water Use and Management in the Bakken Shale Oil Play in North Dakota. Environmental Science and Technology 50, 3275–3282.

Kim, Seongyun, Omur-Ozbek, Pinar, Dhnasekar, Ashwin, Prior, Adam, Carlson, Ken, 2016. Temporal analysis of flowback and produced water composition from shale oil and gas operations: Impact of frac fluid characteristics. Journal of Petroleum Science and Engineering 147, 202–210

Li, Shuifu, Hu, Shouzhi, Xie, Xinong, Lv, Qian, Huang, Xin, & Ye, Jiaren, 2016. Assessment of shale oil potential using a new free hydrocarbon index. International Journal of Coal Geology 156, 74–85

Meegoda, Jay N., Rudy, Samuel, Zou, Zhenting, Agbakpe, Michael. Can Fracking Be Environmentally Acceptable? Journal of Hazardous, Toxic, and Radioactive Waste. 1–11.

Murray, James W. "Limitations of Oil Production to the IPCC Scenarios: The New Realities of US and Global Oil Production." Biophysical Economics and Resources Quality 13, 1–13. 2016

Sangaramoorthy, Thurka, Jamison, Amelia M., Boyle, Meleah D., Payne-Sturges, Devon C., Sapkota, Amir, Milton, Donald K., Wilson, Sacoby M. "Place-based perceptions of the impacts of fracking along the Marcellus Shale." Social Science & Medicine 151, 27–37. 2016.

Sharma, Suresh, Shrestha, Aashish, Mclean, Colleen E. Impact of Global Climate Change on Stream Low Flows in a Hydraulic Fracking Affected Watershed. Journal of Water Resource and Hydraulic Engineering 5, 1–19.

Energy Security in a Dynamic World

Forrest Fulgenzi

The international Energy Agency defines energy security as "the uninterrupted availability of energy sources at an affordable price". This problem affects both developing and developed countries in very different ways, and this can be seen through various aspects of a countries economy and society. As climate change continues to pose a threat to the environment, it can be seen that resources also become constrained and many populations and ecosystems suffer as a result.

The reality of these events continue to magnify over time, which makes the issue of energy security a problem that should be prioritized. Developing countries face a unique situation, as they are experiencing significant population growth, yet are incentivized to continue using fossil fuels as the infrastructure already exists and is cheaper to scale relative to renewable forms of energy. As a result, there is an increased demand for research into how countries can effectively meet the energy demands of the population while sustainably achieving these goals. Developed countries may see energy security much differently, as energy can be used to not only meet the needs of the population, but may also be used as leverage within geopolitics. One country may use their energy supply to garner strategic leverage over another country, which can be utilized to further a country's agenda or regional policy. Through this different route, energy security becomes defined as a much broader topic that involves international relations, economics, and environmental science.

Studies suggest that in order to curb anthropogenic climate change, there needs to be increased cooperation among developed and developing countries to ensure sustainable growth. Increasing cooperation among countries whether it is regional or intercontinental would not only benefit the countries themselves, but would mitigate negative environmental externalities as well. Studies also suggest that there needs to be increased investment within renewable energy infrastructure to ensure sustainable and reliable energy procurement. In the short run, these policies may be expensive and costly, but in the long run, they will be beneficial in that energy will be cheaper and will benefit ecosystems a lot more.

This chapter explores several studies on energy security that delve into both geopolitical theories and economic theories, whose aim is to further ensure sustainable energy procurement at an affordable price while taking into account negative environmental externalities. While there are many different ways to approach energy security, many of the studies examined in this chapter focus on mathematically modeling energy scenarios or on the application of certain energy theories in context of specific countries. A common trend within all of these studies is that a more efficient economy requires implementation of better energy infrastructure with an emphasis on renewable energy. As population growth is continuing to rise, it is imperative that countries attempt to maximize energy investment and allocation so as to maximize the well being of their populations.

Sustainable Energy Security for India: An Assessment of the Energy Supply Sub-System

As one of the world's foremost developing countries, India provides a unique case in which to examine energy security. India defines its sustainable energy security (SES) policy as "provisioning of uninterrupted energy services in an affordable, equitable, efficient, and environmentally friendly manner." According to India's energy security policy, the end goal of any developing country should be to achieve this level of energy security and resource independence. The World Energy Outlook forecasts that India's energy demand will significantly change during the period of 2014–2040 (IEA 2015.), where it will experience a move to the center stage of the world energy system, which will cause a shift in demand. Thus, India needs policies for rapid expansion of energy systems while also looking for a sustainable means to achieve these goals.

Narula *et al.* (2017) provide a new method to assess India's SES that creates a clearer picture of what the country requires to achieve these ends. Narula *et al.* argue that a proper analysis needs to take into account three sub systems of the energy sector: energy supply, energy demand, and energy conversion and distribution. Variables that are taken into account within each sub system include: domestic energy resources, availability of resources, affordability, acceptability, efficiency, and net imports/exports within the country. The researchers then weighted and normalized all of these metrics and then created a composite score to determine an overall SES score.

These composite scores took on a value of anywhere between 0, being the least secure, to a 1, being sustainable and independent. Results indicated that India's SES score was a 0.75, indicating that there is still room for improvement. Results also indicated that the SES index for import of nuclear fuel and for coal is much higher than for domestic nuclear fuel, which suggests efforts to import fuel need to be undertaken to mitigate environmental impacts. Lastly the researchers also suggested moving away from imported natural gas as inadequate infrastructure and limited storage capacity diminished the SES index for natural gas imports. An alternative to this solution would be to invest more in domestic production of natural gas so as to maximize the SES index.

This analysis becomes useful in determining how best to approach energy policy in the future, and gives a glimpse into the specific sub sectors that need to be addressed as well. As a rapidly developing country, the economy is growing with different sub sectors growing at different rates, and this new SES assessment can help give a better idea into which sub sectors need more attention as growth continues.

Forrest Fulgenzi

Illuminating the Policies Affecting Energy Security in Malaysia's Electricity Sector

Over the past few decades, Malaysia has witnessed rapid expansion in its economy and population, spurring it to undergo drastic changes. One of the most significant changes has been within its energy sector. As a country lacking many natural resources, Malaysia has been forced to look abroad to other countries to fulfill its energy needs, leaving it vulnerable to oil and coal price fluctuations. Beckhet and Sahid (2016) analyze current policy implementation that has attempted to solve this energy problem, while proposing a solution for Malaysia's energy future.

Energy security is of great importance to Malaysia, as the electricity sector relies on roughly 90% fossil fuels. To lower their reliance, Malyasia has implemented the National Depletion Policy, and the National Energy Policy, and the Five Fuels Strategy, which have focused on diversifying fossil fuels and introducing renewable energy into its energy portfolio, but much more work that needs to be done.

Beckhet and Sahid argue that Malaysia will need to develop specific strategies to mitigate energy concerns, including diversifying resources of energy supply, reducing carbon content of energy, efficient utilization of energy, and promoting low carbon industries. They indicate that to curb being a net importer, investing in supply domestically both through gas and renewable energy development would be important as well as investing in infrastructure domestically so as to insulate themselves from oil disruptions abroad. Another strategy proposed is implementation of policy that encourages reducing the carbon content of energy. This can take the form of implementation of nuclear power or increased investment in renewable energy. The third strategy is through creating a more efficient energy sector, which mitigates waste and greenhouse gas emissions. The final strategy that the researchers propose is facilitating low carbon industries such as high efficiency products and smart grid system producers, which would promote economic growth and environmental sustainability. These four strategies would be a guiding force into how the country can seek to grow its economy sustainably while being able to meet the energy demands of the country.

Managing the Food, Water, and Energy Nexus for Achieving the Sustainable Development Goals in South Asia

South Asian countries face a number of problems regarding food, energy, and water security. Rasul (2016) analyzes current policy implementation within various South Asian countries and argues that each of these problems cannot be solved in isolation; rather they all interact and contribute to one other.

As a region that is rapidly growing, it is imperative that the peoples' needs are met while sustainably using both renewable and natural resources. Rasul indicates that a main driver for rapid degradation in the environment can be inefficient government policy, which greatly influences development but also has the potential to cause negative impacts. Initially implemented to create food self-sufficiency, subsidies are one example of a policy that creates negative externalities to the environment. Excessive use of the subsidized input contributed to accelerate degradation of natural resources such as land, soil, water, and environmental impacts such as water pollution and biodiversity loss. These impacts are both environmental, and economic because subsidies cut into potential revenue of the energy firms. As a result, policy that is meant to promote one outcome, may inadvertently promote revenue loss and environmental damage in other sectors. Consequently Rasul calls for a more synergistic framework of these three sectors. By taking into account all three sectors when implementing policy, governments can effectively mitigate conflicts and externalities and create effective outcomes.

Rasul suggests a few solutions to streamline more synergistic outcomes. These solutions include strengthening the three sectors by better coordinating policy implementation, so that externalities can better be accounted for. Regulation of unsustainable practice is another method that promotes synergies, as it aims to mitigate water waste. Another solution that Rasul recommends is through the increased investment in infrastructure both for renewables and non-renewables. By investing in water and energy saving technologies, farming practices such as irrigation can become more efficient which has a downstream effect on the economy and the environment as well. The last solution that is recommended is using a more interdisciplinary framework, where decision-making can have a more integrated approach that offers more innovative solutions. These solutions can provide a key component to ensuring energy security in the region while avoiding the harms of a growing population.

An Integrated Examination of Energy Security, Economic Growth and Climate Change Related to Hydropower Expansion in Brazil

With electricity consumption set to increase over 60% by 2030 in developing countries (Kaygusuz 2012), Brazil attempts to meet these energy demands through hydropower. Prado *et al.* (2016) reviews current policy implementation of Brazil's energy security plan, and argues that current policymaking is inefficient to meet energy consumption needs.

Prado *et al.* examines the energy crisis of 2001, which triggered rationing of energy in both California and Brazil. This event elicited a response in the community to become more energy-efficient, but the response among these two actors was drastically different. California changed its demand side of energy consumption prompting energy conservation measures, while Brazil took a policy approach to change its supply side, options to expand production, which increased infrastructure investment as their solution to energy security.

By analyzing current market data and historical trends, Prado *et al.* indicate that the Brazilian Planning Agency drastically overestimates the increased energy demand over the next two decades, and claim that inflated estimates of energy consumption can waste taxpayer dollars by creating a sunk cost of investment, and that excessive investment can lead to unnecessary environmental damage. Hydropower may be an efficient renewable form of energy; however there could be significant environmental impacts since many of the facilities are placed on the Amazon River. The environmental externalities of hydropower placement in the Amazon are poorly understood, but could pose an extreme danger to indigenous groups and to local biodiversity within the region. Consequently, Prado *et al.* argue for a more efficient energy security policy, as the current move to hydropower may actually be a step backwards in environmental sustainability. The researchers argue for a more diversified portfolio of energy sources instead of heavily relying on hydropower. The researchers also call for a better analysis of energy consumption, as the current prediction grossly overestimates demand, which could skew policy making for the future. This review is useful in understanding how Brazil's energy policy may be wasting current resources, and the trajectory in which hydropower is heading under current policymaking.

Comparing Electricity Transitions: A Historical Analysis of Nuclear, Wind and Solar Power in Germany and Japan

Germany and Japan constitute a good case study to examine the evolution of energy security, as both countries started out on a similar path and have diverged greatly over time. Cherp *et al.* (2017) examine this relationship by using a comparative framework that incorporates a technological, economic, political, and sociological perspective to better understand the evolution of energy security between Germany and Japan. Between 1960 to 1990 both Germany and Japan expanded nuclear power to quell their energy insecurities, however the countries' paths diverged in the 90s, when Germany phased out nuclear by investing in wind and solar while Japan focused on increased infrastructure investment in nuclear.

Cherp *et al.* collected statistics from the International Energy Agency (IEA), the International Renewable Energy Agency (IRENA), and the International Atomic Energy Agency (IAEA) to construct a comparative framework that would allow for an understanding of each country's progressive change in energy security. Results suggested that Japan underwent a larger increase in electricity demand, than did Germany's, and it continued to rise, in contrast to Germany's which stagnated over time. Another significant finding was that over time Japan became an energy importer while Germany had a large supply of domestic fossil fuels such as coal. Cherp *et al.* argue that these factors helped define current energy policy, trajectory, and regimes within each country. Results indicate that both countries may have fostered different energy regimes because there was no alternative choice at the time, but they also show that policy implementation may have not always led to a favorable outcome. The researchers argue that in reality, polices may be fragmented and policy makers should be more wary of other variables that are being implemented in tandem with the policy. Cherp *et al.* call for a more interdisciplinary approach to understanding and shaping energy security.

Microalgal Biodiesel: A Possible Solution for India's Energy Security

A common energy problem within India is that there is a growing population and demand for energy that is creating downward pressure on the country's energy supply. As a result, India is pivoting towards new ways of ensuring energy security while meeting its population's energy demand. Sharma and Singh (2017) advocate for a new alternative microalgal biodiesel fuel to be integrated in the energy network, which could help mitigate the need for more energy imports. Sharma and Singh argue that a new fuel solution should not impact or infringe on other sectors such as food or water, and should have minimal environmental consequences. The researchers indicate that the implementation of microalgal biodiesel has a marginal impact on other sectors and has comparatively less Greenhouse Gas (GHG) emissions.

As a developing country, India faces a situation in which it must prioritize poverty reduction, social welfare, and increased living standards, and integrating a new way in which to cut energy emissions may seem daunting. Through their analysis, Sharma and Singh indicate that this new fuel might be efficient, but implementing it on a large scale may not be economically feasible. Large-scale cultivation of microalgae is not only expensive but the commercial technology is

undeveloped. . Singh and Sharma ultimately conclude that more research needs to be done on this type of technology, and argue that
many developed nations should also pursue this research, as it benefits everyone. Through interstate cooperation on the research of new
fuels, countries may be able to pool resources to optimize fuel consumption while mitigating inefficiencies. The researchers argue that
this would be most beneficial to developing countries as it could reduce the negative environmental externalities of industrialization.
These findings are significant in that this poses an interesting thought
experiment that runs counter to India's narrative of energy insecurity.
Sharma and Singh isolate the inefficiencies of rapid industrialization
and urbanization, and propose an alternative way to curb GHG emissions.

Energy Demand, Substitution and Environmental Taxation in Eight Subsectors of the Danish Economy

A response to many European countries moving towards renewable energy initiatives is an increased scrutiny of energy production and consumption. Møller (2016) analyzes data from the Danish
economy spanning more than 55 years to determine which sub sectors in the economy are most subject to volatility. He uses basic economic principles to guide his econometric analysis into energy volatility, and suggests that by using economic incentives such as lower taxes on electricity, governments may be able to induce substitution of
energy inputs among large firms, which could be a way to facilitate
more environmentally friendly fuels. There are eight sub sectors that
account for total industrial energy consumption and economic activity, which is represented by primary, secondary, and tertiary sectors.
These sub sectors include: agriculture, food manufacturing, chemical
manufacturing, machine and vehicle manufacturing, other manufacturing, and construction. By breaking up the industry into eight subsectors, Møller is able to see which areas in the economy are most
susceptible to volatility, and therefore ideal targets for policy implementation.

After running the time series data through regression analysis, the results suggested significant effects among agriculture, machine and vehicle manufacturing, construction, trade and other services, indicating that these five subsectors are susceptible to the most volatile price fluctuations. After this initial discovery, an impulse response experiment was then carried out to investigate whether taxation could induce substitution for electricity by other greener forms of consumption. Results confirmed this hypothesis. This experiment is useful in giving policy makers the ability to directly see how effective specific policy implementation will be in targeting specific subsectors of the economy. The data also provide a new way to estimate potential revenue that may be gleaned through the implementation of targeted taxes.

Energy Security in East Asia Under Climate Mitigation Scenarios in the 21st Century

Japan, China, and South Korea all rely heavily on energy imports to meet their energy needs. Andriosopoulos (2016) investigates how these countries' energy security policies will be shaped under future climate mitigation policies. A computable general equilibrium model was constructed to assess how climate mitigation polices affect economic activity both among energy and non-energy sectors. Two IPCC Representative Concentration Pathways, 4.5 W/m² and 2.6 W/m² were used for the climate change policy scenarios radiative forcing levels in 2100. Three models were generated for both of these scenarios and a third was a reference model, generated without any mitigation policy. Results suggested that energy demand will increase in China, but decrease after 2050 for Japan and Korea. The two policy scenarios estimated primary energy demand to be smaller relative to the reference model, which indicates that reducing total energy demand is required to reduce greenhouse gas (GHG) emissions. Energy structure was also affected within the policy scenarios, with Japan and Korea utilizing more nuclear energy and renewables while all three countries shifted away from fossil fuels. The Herfindahl Index was applied to the three models to determine energy supply diversity.

In the reference scenario absent policy implementation, China's diversity worsens over time while it improves in Korea and increases in Japan during the later half of the century. Among the policy models, diversity increases among all three countries over time, yet the rate at which diversity occurs fluctuates, which indicates that it may not contribute to overall energy security as previously postulated.

This study is novel in that it demonstrates that climate mitigation not only reduces GHG emissions, but policy implementation can facilitate energy security improvements through the diversification of energy. In the short run, these improvements may hamper economic growth for these countries as diversifying energy supplies can be costly, but in the long run it can be beneficial as it reduces energy procurement risks while reducing GHG emissions.

Energy Security in ASEAN: A Quantitative Approach for Sustainable Energy Policy

The ten members of the Southeast Asian Nations (ASEAN) along with China and India are exerting major influence within the global energy sector as they are continuing to grow, yet these countries continue to face significant energy hurdles. Tongsopit *et al.* (2016) investigate the energy security of ASEAN countries using a framework that incorporates the availability, applicability, acceptability, and affordability of energy among these members. Availability refers to the geological existence of fossil fuel energy resources and the degree to which they are being replaced by alternative energy resources. Applicability refers to the ability to use new technologies to ensure efficient use of remaining fossil fuel reserves. Acceptability refers to how an economy or society perceives specific energy resources, both the environmental and societal aspects of energy. Affordability refers to the economic sense of energy, taking into account the price of energy for a series of technologies. These categories were compared against other sets of energy indicators to determine a framework that is best at understanding energy use.

Time series data from 2005–2010 was incorporated from the World Bank and the International Energy Agency (IEA). For each data category, a range of 1–10 was applied, where 1 was the most energy insecure while 10 was the most energy secure. Results suggest that ASEAN countries made little progress and even regressed in terms of establishing energy security under these four variables of interest. The only category in which the countries did not regress was in the applicability category, in which the countries significantly increased.

Overall, Tongsopit *et al.* advocate for an increased development of renewable energy and energy efficient technologies, as these would move ASEAN countries in a positive direction towards sustainable energy goals. The findings suggest that energy is becoming more expensive, and the researchers propose cooperation among countries as a solution to this problem, alleviating some of the demand and would lower the prices of electricity. Utilizing a framework that incorporates various categories is also beneficial, as it can motivate ASEAN countries to participate in regional goals that have specific outcomes, which could facilitate regional cooperation and strategy. ASEAN countries are witnessing a decline in energy security among as a result of economic growth, population growth, and declining fossil fuel reserves. An increase in overall security would requires improvement in all four dimensions of energy security, and researchers suggest need for policy action to coordinate energy planning efforts on a regional scale.

Designing Energy Supply Chains: Dynamic Models for Energy Security and Economic Prosperity

Many developing countries face energy deficiencies; they have plenty of natural resources, yet lack the financial ability to use these resources optimally. Mun *et al.* (2017) analyzes current coal supply chains in Pakistan, and creates a new mathematical model that optimizes the supply chain while taking into account developing countries' economic constraints.

Mun *et al.* create a new mathematical model that incorporates an economy–dependent public budget to capture and optimize the cost and time trade-offs in a developing countries' energy supply chain. Unlike a developed country, a developing country is severely limited by its economy, which prevents efficient capital investment in its energy infrastructure. The researchers create a multi-period mixed integer linear program model that minimizes energy gaps within the supply chain. Results from the model suggest building power plants near demand zones and obtaining coal from new reserves first. They also advocate a tiered strategy, with the government exploring smaller reserves initially and moving to larger reserves over time. This would ensure efficient extraction of resources while being able to build up reserves through the creation of newer plants. Through this tiered strategy, inefficiencies can be minimized, as this model seeks to isolate limiting factors in the supply chain and fix them.

The researchers also argue that even though their findings advocate a more efficient use of resources, measures by the government should still be taken to account for environmental externalities in the short run, and a more diversified energy supply chain in the long run. The authors also argue that environmental policies should be implemented in tandem with these supply chain reforms, to offset the environmental externalities of increased production of coal.

Transatlantic Energy Security: On Different Pathways?

As US energy production continues to rise and EU production decreases, Beyer and Fischer (2016) evaluate the current US and EU approach to energy security and how it will be shaped by politics and diplomacy. The EU is experiencing an increase in foreign energy imports coupled with a gradual decline in domestic production of oil and gas. Having to rely on external countries for sources of energy becomes dangerous, as it leaves one open to oil disruptions and price fluctuations. In response to this energy import problem, the EU created an energy union to address problems such as energy security and energy dependence, yet has failed to answer issues such as implementation of renewables and of energy efficiency, hampering the effectiveness of the union. It appears that the EU and the US are moving in quite opposite directions, which only magnifies the importance of transatlantic cooperation. Beyer and Fischer argue that the US is far from being an energy superpower, as it doesn't have the infrastructure in place to become a sole exporting country nor the ability to rely on its own supply for the time being. As a result, it is within the best interest of both parties to cooperate in regards to trade.

As the US exports more energy, Beyer and Fischer argue that the US will leverage this as another tool for diplomacy and foreign policy. Examples of this include supplying areas of strategic interest with Liquid Natural Gas (LNG) or state intervention in the determination of gas pipelines. For the EU, transatlantic cooperation will not be perceived as an opportunity to garner leverage, rather it will function as an opportunity to acquire secure access to energy imports while being able to wean imports from of countries such as Russia which have been a significant share of gas imports for the EU. These two entities are experiencing significant differences in regards to energy production than they did decades ago, and as a result of this change in energy cooperation, it is more significant than ever to find a way to work together to achieve the best outcome for everyone.

Conclusions

By incorporating these various aspects of energy security, policy makers can better make decisions regarding their population and the environment. Through increased investments in renewables and better modifications within one's supply chain, a country can enjoy the benefits of a fossil fuel-based economy with renewable energy. Energy security will continue to be a pressing issue regardless of what type of economy a country has, and as a result, there should be increased research into this broad field, so that policymakers can make optimal decisions. Energy security can be seen as a prerequisite to tackling issues such as global warming and environmental degradation, as having the appropriate infrastructure is key to managing these issues.

References Cited

Andriosopoulos, K. (2016). Energy security in East Asia under climate mitigation scenarios in the 21st century. *Omega, 59,* 60‑71.

Bekhet, H. A., & Sahid, E. J. M. (2016). Illuminating the Policies Affecting Energy Security in Malaysia's Electricity Sector. World Academy of Science, Engineering and Technology, International Journal of Social, Behavioral, Educational, Economic, Business and Industrial Engineering, 10, 1164-1169.

Beyer, A., & Fischer, S. (2016). Transatlantic Energy Security: On Different Pathways?.

Cherp, Aleh, *et al.* "Comparing electricity transitions: A historical analysis of nuclear, wind and solar power in Germany and Japan." *Energy Policy* 101 (2017): 612‑628. https://www.researchgate.net/profile/Mehdi_Rasti2/publication/283723396_Energy_demands_and_renewable_energy_resources_in_the_Middle_East/links/56454a4e08ae54697fb86a0a.pdf

IEA (International Energy Agency), 2015. World Energy Outlook 2015. Paris: OECD/IEA

Kaygusuz, K. (2012). Energy for sustainable development: A case of developing countries. Renewable and Sustainable Energy Reviews, 16, 1116‑1126.

Møller, N. F. (2017). Energy demand, substitution and environmental taxation: An econometric analysis of eight subsectors of the Danish economy. Energy Economics, 61, 97–109.

Narula, K., Reddy, B. S., Pachauri, S., & Dev, S. M. (2017). Sustainable energy security for India: An assessment of the energy supply sub-system. Energy Policy, 103, 127–144.

Prado, F. A., Athayde, S., Mossa, J., Bohlman, S., Leite, F., & Oliver-Smith, A. (2016). How much is enough? An integrated examination of energy security, economic growth and climate change related to hydropower expansion in Brazil. Renewable and Sustainable Energy Reviews, 53, 1132–1136.

Rafique, R., Mun, K. G., & Zhao, Y. (2017). Designing Energy Supply Chains: Dynamic Models for Energy Security and Economic Prosperity. Production and Operations Management.

Rasul, G., 2016. Managing the food, water, and energy nexus for achieving the

Sharma, Y. C., & Singh, V. (2017). Microalgal biodiesel: A possible solution for India's energy security. Renewable and Sustainable Energy Reviews, 67, 72-88.

Sustainable Development Goals in South Asia. Environmental Development 18, 14–25.

Tongsopit, S., Kittner, N., Chang, Y., Aksornkij, A., & Wangjiraniran, W. (2016). Energy security in ASEAN: a quantitative approach for sustainable energy policy. Energy policy, 90, 60–72.

Battery Technology and Green Energy in the Transportation Sector

Sloan Cinelli

According to the Intergovernmental Panel on Climate Change, average global temperatures have warmed 1.53°F from 1880 to 2012. Atmospheric carbon dioxide concentrations have also increased by 40% since pre-industrial times, primarily in part by fossil fuel emissions. The transport sector itself represents 23% of global CO_2 emissions, only to increase with our growing population (World Bank, 2016).

A tangible dream of the 21st century is to supply power for personal travel vehicles through efficient, green battery technology. Today, energy storage and usage rely more heavily on natural energy harvested from their environment than ever before. We have already found alternative ways of powering vehicles, but in order to do it on a wide scale, we need more efficient batteries, battery-charging technology, and energy storage.

In 1800, Italian physicist Alessandro Volta published results of an experiment he named the Voltaic pile. This stack of zinc and copper is known as the first electrical battery. In 1891, Nikola Tesla designed an electrical resonant transformer circuit to produce low-current, high-frequency alternating-current high-voltage electricity. In 1991, Sony introduced the lithium-ion battery, having the highest energy density and slowest loss of charge in the commercial market. Even in recent news, teams like Petibon *et al.* (2016) are altering the electrolyte envelope, increasing Li-ion energy density by at least 10%.

Christen *et al.* (2017) are verifying cooling strategies to extend the lifetime of the cells. For the past two centuries, batteries have increasingly become smaller, more powerful, and more efficient, opening the space for green energy development. For batteries to play a part in decelerating global climate change, they need to be produced and recharged in a renewable way.

For example, driving an electric vehicle does not necessarily decrease your carbon footprint. Ellingsen *et al.* (2016) determined that charging the electric vehicles with coal-based electricity made the electric vehicles lifecycle emissions 12%–31% higher compared to conventional combustion engine cars. Some companies have started utilizing green electricity, generated from natural and renewable energy sources, having less of an impact on the environment than fossil fuels. The same authors concluded that electric vehicles charged with wind-based electricity had a 66%–70% reduction in the carbon emissions than their combustion counterparts. Usage conditions, battery efficiency, and storage solutions all contribute to the size of your vehicle's carbon footprint.

This chapter discusses examples, precursors, and consequences of these contributions. Child *et al.* (2016) discussed the development of necessary renewable energy systems to close the gap between variability of energy storage and stability of energy supply, while Christen *et al.* (2017) reviewed the internal resistance of batteries in order to peak the performance of variable energy harvesting communication systems. Petibon *et al.* and Ellingsen *et al.* tested battery technology in order to increase performance in the electric vehicles that Gibson *et al.* (2016) and Ajanovic *et al.* (2016) discussed. The clean and green energy revolution is a group effort, and with innovation and community, we can transition into a cleaner and more secure energetic future.

Harvesting, Storing, and Using Naturally Sourced Energy to Power Communication Transmitters

Today, energy storage and usage rely more heavily on natural energy harvested from the environment. These systems, including solar power, wind energy, or salinity gradients, are becoming more popular due to their renewable nature. However, the power that is generated from using these natural sources fluctuates with time, sometimes randomly. In the case of solar cells, power attained in one unit could range from 1 μW to 100 mW depending on sunlight, a 100,000 timescale difference. Hence, energy supply from these systems are much more variable than in conventional systems.

Bhat *et al.* (2017) attempt to fundamentally change the way harvested energy is stored and used in batteries with different efficiencies. Whenever naturally harvested power is lower than the power required for system operation, a system cannot run from that source alone. First, energy must be stored in a battery, then simultaneously drawn from the battery and the natural source, enabling the system to run from the combined power.

In order to accomplish this feat, Bhat *et al.* considered a low-power wireless transmitter, powered entirely by a naturally harvested source, that is equipped with a battery with different capacity constraints. In this paper, they develop novel policies for managing the battery charging and discharging schedules in the transmitter to promote energy efficiency. This dual-path energy harvesting system contains a power splitter, battery, power combiner, and transmitter.

First, the power splitter instantaneously divides the harvested power in order to charge the battery and power the transmitter. The power combiner then combines power directly from the natural source and battery. The transmitter consumes energy to power the circuit, but does not consume energy if it is not transmitting data. Using this system, Bhat *et al.* then input batteries of different resistances. They found that different internal resistances considerably inhibit the energy redistribution, reducing the average communication rate of the transmitter. With the efficiency data, using batteries of different resistances, they derived compact expressions for optimal time and power splitting ratios. These ratios determine how much power should be sent to charge the battery, versus how much should directly power the transmitter. Using these models of charge/discharge efficiencies as functions of the internal resistance of the battery, transmission systems of this sort can be run optimally.

Improving Technology of Li-Ion Cells for Rechargeable Batteries

The lithium-ion battery is the power source for most modern electric vehicles. Each battery is made up of many smaller units, called cells. The electrical current reaches these cells via conductive surfaces, including aluminum and copper. There is a positive electrode, the cathode, and a negative electrode, the anode. The battery is filled with a transport medium, the electrolyte, so the lithium ions carrying the battery's charge can flow freely from one electrode to the other. This electrolyte solution needs to be extremely pure in order to ensure efficient charging and discharging.

Virtually every lithium ion cell produced today uses ethylene carbonate (EC), and most battery scientists believe it is essential. Petibon *et al.* (2016) tested electrolyte systems other than this within lithiumi-ion battery cells. Surprisingly, totally removing all ethylene carbonate from typical organic carbonate-based electrolytes and adding small amounts of electrolyte additives creates cells that are better than those containing ethylene carbonate. Petibon *et al.* (2016) used different surface coatings, electrolyte additives, and new solvent systems, and the impact was substantial.

Petibon *et al.* tested the conductivity, self-heating rate, volume change, discharge capacity, and polarization growth of cells with differing electrolyte solutions. They found that the removal of ethylene carbonate has been shown to enhance high voltage performance of cells both at room temperature and high temperature. Also, the addition of a co-additive helps to lower the polarization growth during high voltage cycling, improving the safety. Lowering the polarization decreases the mechanical side effects, where isolating barriers develop at the interface between the electrolyte and the electrode. Overall, it is shown that several compounds are able to replace the EC electrolyte, including vinylene carbonate and ethyl methyl carbonate. These results clearly show that large amounts of EC are not needed, and are actually detrimental to the cycle of cells operated with a high voltage.

Testing Heat Flux in Lithium-Ion Cells to Extend Lifetime of Batteries

Reliability and length of life are essential factors for the success of a battery system. In order to maintain peak electrical performance and avoid degradation, the battery cell has to be kept within a narrow temperature range throughout its operational lifetime. In cool temperatures, lithium plating is observed, but in elevated temperatures, solid-electrolyte interphase growth occurs, leading to capacity fade, increase in resistance, and sometimes thermal runaway. If there is a temperature gradient across the battery, the behavior of the cell is dictated by the higher temperature region. Christen *et al.* (2017) tested prismatic lithium-ion batteries, modeling the thermal properties of a battery cell, and verifying different cooling strategies.

Christen *et al.* created a testing mechanism with the ability to acquire and control heat flux and temperature distribution on the surface of battery cells. The experiment was conducted on prismatic cells—thin and light batteries used in most mobile phones. The high energy density lithium manganese oxide cells had physical dimensions of 175 mm × 125 mm × 45 mm. The researchers surrounded one side of each cell with 87 uniformly distributed temperature/heat flux sensor (THFS) units, connected a heat sink to the other side, and locally measured and controlled temperature and heat flux.

First, Christen *et al.* kept all of the cells at 25°C throughout the experiment. Then, they kept the cells in constant heat flux, a more realistic reproduction of the life of a battery. Meanwhile, each cell was cycled between 3.5 and 4 V. The internal temperature gradients built up at the beginning of the experiment, which led to the exponential behavior of the heat flux measurements. As flux measures the heat transferred per unit area per unit time, the first pass of current through the electrolyte caused this exponential change.

Through applying constant temperature to all THFS units at the same time, the heat generation of the battery cell was quantified and localized. Christen *et al.* used Matlab to process the measurements from the THFS units. They found that the heat flux density on the side faces and bottom side of the cell were close to zero; the main part of the heat was almost unvaryingly released through the large front and rear side; and the largest heat flux density was measured in close proximity to the positive terminal of the battery cell.

Usually, electrical conductors exhibit a low thermal resistance and generally good heat dissipation. Therefore, the low heat flux density on the side faces was a surprising result. Also, the asymmetry between the two electrical terminals was an interesting find. In order to verify a cooling strategy, Christen *et al.* performed two follow-up experiments. The most promising locations for cooling the battery cell seemed to be in proximity to the electrical terminals or in the center of the front/rear face. This assumption was made because the dissipated heat flux density was the highest in those areas.

For the first experiment, Christen *et al.* attached seven THFS units to the bottom side of a battery cell to keep it at a constant temperature of 25°C. The second experiment consisted of eight THFS units operating in the temperature-controlled mode, with four units on each terminal. When Christen *et al.* chose a more appropriate location, the terminals, for the active cooling of the battery cell, the surface temperature gradient was reduced by 23%. This is expected to help the cells maintain peak electrical performance. As of now, work is in progress to miniaturize THFS units, so they can be applied to cylindrical cells. Work like this is necessary in order to improve the reliability of entire battery systems.

Using Battery Technology to Increase Viability of Electric Vehicles

In the early 1900s, the number of electric vehicles was almost double that of gasoline powered cars. Today, electric vehicles are far from the consumer vehicle of choice because of high purchase cost, limited driving range, long charging times, and poor durability of the battery. Gibson *et al.* (2016) used technology forecasting with data envelopment analysis to compare future battery performance characteristics in electric vehicles with future performance goals established by the United States Department of Energy. The authors concluded that a new battery technology must be developed, as current incremental improvements in technology will leave electric vehicles considerably short of performance specifications from the DOE.

First, Gibson *et al.* reviewed electric vehicle and battery literature to understand cost concerns. Each battery is characterized by a power-to-weight ratio and an energy-to-weight ratio, measuring the performance of the engine. A balance needs to be maintained between these values because a sharp power increase will cause a decrease in energy. Lithium-ion batteries are most common, as the lightweight properties of lithium result in a cell voltage between 3.3 and 4.3 V; they can store more energy per mass and volume than competing battery types.

Analysts reported the cost of the battery to be approximately 50% of the cost of the vehicle in 2010 and about 30% in 2012. The Office of Energy Efficiency and Renewable Energy recently released funding with the intent of meeting new performance targets, including range, power, and life cycle. This office in the Department of Energy has been chartered to advance the development of batteries and other energy storage devices to enable a large market penetration of electric vehicles.

Gibson *et al.* then used web search results to identify key organizations by using sheer number of publications. The top organizations were identified from the patents for their technology, and their current focus of research and development of the associated technologies was verified online. Government funding for battery research and development was identified in companies in the USA, China, Japan, and South Korea. Further, Japan and South Korea held an 80% share of global production of advanced Li-ion batteries in 2010.

Technology forecasting using data envelopment analysis was employed to compare each organization. These comparisons included differences in battery technology, and the level of research or production each assignee was in. The final, output-oriented model took battery-weight as an input, and output what improvements can be realized in price, MPGe, and acceleration. The authors assumed that the focus of battery technology research and development was to increase the range and acceleration of the electric vehicle while maintaining the same battery weight.

Gibson *et al.* found that there was an average rate of change in battery output per year of approximately 3%, despite the large investments in electric vehicle battery technology companies. Through observing changes in the inputs and outputs of the technology forecasting model over the 16-year period of available data, there did not seem to be any major improvements at all. The impact of the first Tesla model S battery increased the mean standard deviation from 2012 to 2012.25, but it is an outlier due to its energy capacity that far exceeded the other batteries.

In conclusion, for electric vehicles to become mainstream and meet the Department of Energy's goals by 2020, battery performance must improve at a faster rate. The incremental improvements in the battery technologies studied will leave electric vehicles considerably short of the DOE's performance specification for vehicle range. Gibson *et al.* posit that more emphasis should be placed on finding new technologies that can safely operate at higher temperatures. In order to understand the intricacies of technological advancement, they state that it may be beneficial to look at earlier stages of research and development.

Assessing the Greenhouse Gas Emissions of Modern Electric Vehicles

Human passenger vehicles consume about one-fourth of global primary oil. There are approximately one billion of these vehicles on earth, and that number is expected to double in the next few decades. With the increasing popularity of electric vehicles to counteract greenhouse gas emissions, Ellingsen *et al.* (2016) investigated their environmental impact. The authors studied the effect of increasing battery size and driving range on lifecycle greenhouse gas emissions.

The authors modeled four sizes of electric vehicles to establish essential parameters for the study. Using electric vehicles powered by Li-ion batteries, they collected data from mini cars, medium cars, large cars, and luxury cars. Using the same four sizes, they researched emission data for combustion cars. For conventional combustions vehicles, weight and fuel consumption are strongly coupled. Ellingsen *et al.* plotted the energy requirement as a function of curb weight for electric vehicles and determined an increase of 5.6 Wh km^{-1} per additional 100 kilograms.

The authors synthesized and adapted detailed inventories on the electric vehicles in order to understand each lifecycle phase. These stages include battery production, vehicle production, use, and end-of-life treatment. The vast majority of batteries are produced in East Asia, while most vehicle production and assembly takes place in Germany. The authors quantified the gas emissions from the production phases by inventorying which metalloids were used and in what quantity.

For the electric vehicle use phase, Ellingsen *et al.* made two key assumptions; they assumed a lifetime of 12 years and a yearly mileage of 15,000 km, resulting in a total of 180,000 km. The total mileage and energy requirement of each type of vehicle were multiplied to find the total electricity requirement. Attempting to model electricity generation mix, the authors assumed an average of 521 g CO_2 kWh^{-1}.

For the end-of-life treatment, emissions depended on the size of the battery. Independent of battery size, each disposal delivered lithium, manganese, aluminum, and a liquid alloy containing copper, iron, cobalt, and nickel. The environmental impacts were calculated in terms of ton carbon dioxide equivalents per vehicle over a lifetime of 180,000 km.

Ellingsen *et al.* found that the electric vehicle production phase was more environmentally intensive than that of the conventional combustion vehicles, but electric vehicles had lower use-phase emissions. The electric and combustion cars broke even between 44,000 km and 70,000 km, but the larger the electric vehicle, the sooner it made up for the higher production impact. For both types of vehicles, the use phase was responsible for the majority of greenhouse gas emission, whether it was directly through fuel combustion or indirectly through electricity production.

Although, the use-phase cost in electric vehicles is strongly determined by the way electricity is generated. Electricity generated from natural gas had 12%–21% lower impact than their combustion counterparts, and wind-based electricity had a 66%–70% reduction. Regardless of powertrain configuration, decreasing the size of vehicle will decrease greenhouse gas emissions.

In conclusion, electric vehicles do not always have a lower lifecycle impact than combustion engine vehicles. The benefits increase with distance driven and length of lifetime, but that is strongly dependent on the size of the vehicle and the manner in which electricity was generated. As technology and electric vehicle innovation continues to develop, it is important to reassess environmental impacts in order to evade potential pitfalls.

Increasing the Use of Electric Vehicles in Urban Areas to Reduce Gas Emissions

The transportation sector is one of the major contributors to global energy consumption and greenhouse gas emissions. The European energy and transport policy declares the need for a more environmentally friendly energy and transport system. According to the EU's Energy and Climate Change Packages, a 10% share of renewable energy should be achieved in the transport sector by 2020, the use of combustion engine cars in urban transport should be reduced by 50% by 2030, and completely phased out by 2050. Ajanovic and Haas (2016) identify the major impact factors behind the broader dissemination of electric vehicles in urban areas and compare different countries which are currently active in distribution of electric vehicles.

First, the authors compared the number of electric vehicles. Although the absolute number is highest in Los Angeles, Shanghai, and Vienna, the electric vehicle densities are low in comparison to the total number of registered vehicles. Per capita, the number is highest in Oslo, with almost eleven electric vehicles per thousand inhabitants. Oslo is followed by Rotterdam and Amsterdam.

Next, the authors reviewed charging infrastructures. The number of electric vehicles per charging station ranged from 0.3 to 19. This ratio is highest in Los Angeles, Oslo, and Rotterdam. In comparison, the number of charging stations in Stockholm is more than three times higher than the number of electric vehicles.

After investigating different indicators such as the relation between electric vehicles per capita and GDP, Ajanovic and Haas highlighted the current policies and measures implemented by many local and national governments worldwide to promote the use of electric vehicles. Although most are monetary exemptions or reductions from road taxes, including annual circulation tax, registration tax, or fuel consumption tax, there are also different, non-monetary rewards. These include free parking spaces, the possibility for electric vehicle drivers to use bus lanes, and a wide availability of fast charging stations. These different policies are used to increase the attractiveness of the vehicles, and the majority of them are implemented in urban areas.

In Norway, electric vehicles are exempt from registration tax, value-added tax, annual car tax, and road toll and congestion charges. Drivers of electric vehicles in Oslo have access to bus lanes, free parking spaces, and a good public charging network with about 10,000 charging stations. At the same time, the gasoline price in Norway is high, and electricity price is relatively low. Measures like these have successfully driven Oslo to having the highest number of electric vehicles per capita.

After analyzing the different policies and measures, Ajanovic and Haas found that the economic attractiveness of electric vehicles depended on policies implemented, travel activity and the ratio of fossil fuel to electricity prices. Subsidies and tax incentives for electric vehicles are an important tool for dissemination, but according to Ajanovic and Haas electric vehicles make sense only if it is ensured that a major share of electricity they use is generated from renewable sources.

Understanding Energy Consumption of Electric Vehicles

Road transport contributes about one-fifth of total carbon dioxide emissions in the European Union, and private vehicles alone are responsible for 12%. The EU has targeted a 60% emission reduction by 2050, so electric vehicles will play an important role in reducing greenhouse gas emissions in the near future. Li *et al.* (2016) reviewed all possible factors which may influence the energy consumption of electric vehicles, explored the statistical significance of selected factors, and created a binary model for predicting consumption.

After reviewing existing models for consumption, Li *et al.* grouped all factors into 6 categories: technology and vehicle, artificial environment, natural environment, driver, travel type, and measurement. There had been no models derived from real-world data to depict electric vehicle performance in a specific region, so this study presented the impacts of topography, infrastructure, traffic, climate, and their interactions. The authors applied Design of Experiment methodology to explore the statistical significance of each selected factor.

In order to simulate the real world usage phase, the Nissan LEAF was tested in the metro area of Sydney, Australia. The car weighs 1521 kg, with a drag coefficient of 0.28 with closed windows. The lithium-ion battery voltage is 345 V with a maximum storage of energy 24 kWh. A 4.17 km trial route was designated, and 25 runs were conducted. The experimental setup consisted of the vehicle, a controller area network on-board diagnostic Bluetooth scanner, and an Android device for data collection.

The results Li *et al.* found were not unexpected. Increased traffic on a hilly route resulted in more acceleration on a slope, causing higher energy demand. The combined effect of traffic and topography was shown significant, as well as traffic and climate. However, the interaction of topography and climate was not significant.

Using the significance level of tested factors and their interdependency, the researchers developed a binary model. Li *et al.* derived a linear equation defining the energy consumption in Wh/km.

$$y = 108.19 + 41.22 * D + 39.34 * C + 22.4 * B + 17.88 * A$$

A, B, C and D refer to the infrastructure, traffic, topography and climate respectively. The climate and topography had a greater impact on the energy consumption than the other factors. The binary model confirmed these findings, and it can be used for rough estimation. All of the tested factors showed significant impacts on energy consumption of electric vehicles, with some interactions between factors causing a secondary effect. This shows that specific usage conditions are extremely important when attempting to quantify the energy consumption of an electric vehicle.

Energy Storage Solutions in a 100% Renewable Finnish Energy System

The supply and demand of energy services vary over time and space, making variability and uncertainty inherent qualities of energy systems, particularly in the case of renewable energy. Increases in the development of renewable energy systems is needed to close the gap between variability of energy storage and stability of energy supply.

Using hourly data analysis, Child *et al.* (2016) determined the roles of various energy storage systems in Finland. Electricity and heat from storage represented 15% of end-user demand, while thermal storage was 4% of end-user heat demand. For the power sector, 21% of demand was satisfied by electricity storage, while 87% was from vehicle-to-grid connections. Grid gas storage satisfied only 26% of gas demand. These data proved that energy storage solutions will be very important to the future of green energy in Finland.

Finland has committed to an 80–95% reduction in greenhouse gas emissions by 2050. Variation in renewable energy generation is typical of a country at this latitude. After analyzing current data, Child *et al.* developed a 100% renewable energy scenario for

Finland using the EnergyPLAN modelling tool. The EnergyPLAN advanced energy system analysis is a computer model that deciphers the best 100% renewable energy scenarios. This accounts for biomass resource availability, and cost-competition.

In 2015, 105 TWh of electricty and 65.3 TWh of heat were consumed. Both the annual electricity and heat that were derived from stored gas were calculated according to the specific summation equations. In order to determine the usage of solar and wind energy that would constitute a transition to 100% renewable energy, the sum of each energy production category was divided by total supply of electricity from all sources.

Child *et al.* found that solar and wind power will make roughly 60% contribution to final energy consumption, and 70% of total electricity generation. Approximately 47% of the renewable energy will be utilized directly, with 51% going to storage. They determined that electricity storage devices will be needed once 50% of power demand is met with variable renewable energy, and seasonal storage devices will be needed after more than 80% of electricity demand is met by renewable energy. Variable renewable energy and energy storage solutions play the most significant role in a 100% renewable energy solution for Finland, so models like this contribute very heavily towards the 2050 goal.

Conclusions

We have overloaded our atmosphere with greenhouse gasses, resulting in global climate change. Carbon dioxide is putting us at risk of irreversible change, and CO_2 emissions will only increase as the human population continues to grow. Innovation and exploration within the green energy sector are necessary to improve energy security on our planet. The road to a cleaner and more secure future is paved with more efficient energy storage and usage technology. This is most feasibly acquired by utilizing rechargeable battery technology, with electricity created by renewable energy sources. Proliferating our use of electric vehicles will decrease our carbon footprint, and increase our chances of having secure and renewable energy sources in the future.

References Cited

Ajanovic, A., and Haas, R., 2016. Dissemination of electric vehicles in urban areas: Major factors for success. Energy 115, 1451–1458

Bhat, R.V., Motani, M. and Lim, T.J., 2017. Energy Harvesting Communication Using Finite-Capacity Batteries with Internal Resistance. *arXiv preprint arXiv:1701.02444.*

Child, M., & Breyer, C., 2016. The role of energy storage solutions in a 100% renewable Finnish energy system. Energy Procedia 99, 25–34

Christen, R., Rizzo, G., Gadola, A., & Stock, M., 2017. Test Method for Thermal Characterization of Li-Ion Cells and Verification of Cooling Concepts. Batteries 3, 3

Ellingsen, L., Singh, B., & Strømman, A. H., 2016. The size and range effect: lifecycle greenhouse gas emissions of electric vehicles. Environmental Research Letters 11, 5 http://iopscience.iop.org/1748-9326/11/5/054010/

Gibson, E., van Blommestein, K., Kim, J., & Garces, T., 2016. Forecasting the electric transformation in transportation: the role of battery technology performance. Technology Analysis & Strategic Management 1–18 http://dx.doi.org/10.1080/09537325.2016.1269886

Li, W., Stanula, P., Egede, P., Kara, S., & Herrmann, C., 2016. Determining the main factors influencing the energy consumption of electric vehicles in the usage phase. CIRP 48, 352–357

Petibon, R., Xia, J., Ma, L., Bauer, M. K. G., Nelson, K. J., Dahn, J. R., 2016. Electrolyte Systems for High Voltage Li-Ion Cells. Journal of The Electrochemical Society 163, 13. http://jes.ecsdl.org/content/163/13/A2571.full

Advancing Solar Technology

Michael Crowley

As local and global climates continue to change, the development of renewable, sustainable and cost-effective energy options is of utmost importance. Solar power is one renewable energy source that has shown significant success, and continued progress, in meeting our energetic needs. Over the past few years, the solar industry has been racing towards fabricating highly cost-effective and, most of all, highly efficient solar cells. To do this, researchers have employed a variation of techniques to confront some of the most complicated problems facing modern solar cells.

In the past year, novel research has given extensive insight into many of the crucial problems hindering cell efficiency, stability, and longevity. One key area of growth in 2016 was the diversification, and success, of doping techniques. These new techniques aimed to reduce grain boundaries (GB) and recombination in the films of solar cells. Limiting, and reducing the size of, GBs creates a more uniform film surface, giving all solar cells the ability to immediately become more efficient by reducing recombination and energetic loss. Ion implant technology, emitter passivation, and post-annealing practices have all been proven to provide more uniform surface areas and increase efficiencies in multiple different cell types.

Commercial cells have also seen major improvements. Researchers all over the world have completed extensive research projects into perovskite solar cells as a potential alternative to the classic silicon commercial cell. Perovskite cells utilize perovskite-structured compounds as light harvesting layers. Perovskite compounds are extremely simple to manufacture and are thus cheaper to produce, which is of vital importance when discussing the further commercialization of solar cells. Efficiencies of perovskite film cells have reached well over 20% while maintaining one of the lowest fabrication costs in the industry. Silicon cells also continued their steady progress, with a study on organic layer passivation (Shinde *et al.* 2016) particularly standing out.

Although much of the recent research has focused on perovskite cells, other cell types have enjoyed consistent growth in the past year. Extended works into stacked solar cells for space applications and organic solar cells were published and highlight the diversity and magnitude of the work being done in the solar industry. All cell types bring different strengths and weaknesses to the industry as a whole. Continued progress and investment into all of them will facilitate continued success of solar power in to the future.

Solar power is rapidly becoming a viable energy option for homeowners, businesses, and governmental projects. It is being used around the globe and efficiencies are constantly rising while costs are being cut. In order to fully understand the magnitude of the recent research, detailed reviews of eight papers from this past year are included below.

Ion Implant Technology and ITS Applications to Solar Cells

The once assumed idea, that photovoltaic (PV) solar technologies could not be cost effective and energetically viable, is slowly crumbling. Relatively low-cost and powerful solar cells are now produced in large quantities throughout the world, making the levelized cost of electricity (LCOE) of these cells competitive with other energy-related technologies. A similar thought process was assumed when discussing the possibilities for ion implant technology in solar applications. This assumption is slowly being proven wrong too.

Ion implanting technology is slowly becoming a favorable method for doping formation in semiconductors (silicon is commonly used in present PV cells). By using this technique, higher-efficiency solar cells can be mass-produced. Two hurdles must be faced in order to make this technology more accessible: high costs and low throughput. These hurdles have been faced most head on by the semiconductor company, Kingstone. By developing a simpler ion implanter, Kingstone believes they can cut 90% of their current operations cost while producing a very high throughput (>1500 wafers/hour).

Their researchers in Shanghai looked into the possibilities for ion implanted solar cells and the ability for ion implanters to become low cost but high throughput options. If doping is needed more than once, ion implantation has clear advantages. Fewer steps are required in the workflow and efficiency loss at each step is limited. Ion implementation, which produces cells with an average of ~21% efficiency, avoids the <0.1% efficiency losses often found when using laser edge isolation (eg. plasma etching), and this number increases as surface size increases. For standard p type cells, ion implantation shows similar performance with conventional techniques.

Ion implantation allows for the ability to use uniquely single sided doping. This enables easier and simpler production in n-PERT and IBC cells and will result in higher efficiencies in both mediums. In a forming one specific type of n-PERT cell, poly-Si, ion implantation will boost average efficiency to over 22%. In IBC cells, that number will be raised to nearly 22.5%, confirmed by a third party test. Commonly used in the semiconductor industry, ion implantation, and the range of possibilities that comes with the technique, is now primed to be fully utilized in the solar industry.

Emitter Passivation of Silicon Solar Cells via Organic Coating at Room Temperature

Silicon photovoltaics are a proven sustainable energy solution and take up 90% of the solar cell market. Silicon cells have many practical advantages associated with them, including high cell efficiency, stability and longevity. They are also extremely cost effective. The cost of mass produced silicon cells has dropped below $1/W, in some cases as low as $0.3/W. At the same time, efficiency continues to increase to 20%. The biggest hindrance to further increases in efficiency are high rates of recombination at the surface of the cells, which is what Shinde *et al.* (2016) have been working on.

In order to lower the rates of recombination, the technique of passivation has been employed. SiN_x and SiO_2 compounds have been used for passivation in the past, although the desired surface passivation is accomplished, these compounds require high process temperatures (300°–1000°C). At these temperatures, the properties of the silicon crystalline structure are affected. If the temperatures are reduced, efficiency and longevity are expected to increase.

To combat high process temperatures, new techniques have been presented. It has been shown that passivation can be achieved by using Si-O and Si-H and organic passivation. Shinde *et al.* (2016), look at passivation of n-type emitter by organic cover layer Oleylamine (OLA). This passivation technique will increase efficiency and has the ability to be processed at room temperature.

When comparing basic, standard silicon cells to cells passivated with OLA, many interesting results were found. Uncoated emitters of standard cells register reflection values between 27–30% for the 300–800 nm range. When the same cells are coated with OLA, the reflection efficiency increases. This result shows that OLA is achieving the goal of serving as a passivating layer. It also rules out the possibility that this layer acts as an anti-reflection layer.

When analyzing cell efficiency, OLA coating proves successful as well. Without the OLA coating, cell efficiency was registered at 13.34%. With the OLA coating, that value shoots up to 15.20%. This again shows the ability for OLA coating to successfully passivate the cell surface. Finally, by looking at Raman measurements, bonding with the silicon is clearly achieved. Raman measurements observe vibrational, rotational and low frequency processes in a system.

All of these results are achieved at room temperature and this makes the finding very significant. OLA coating has a huge potential to significantly reduce thermal requirements for the fabrication of solar cells and potentially reduce payback time frames for photovoltaic systems.

High Efficiency and Radiation Resistant InGaP/GaAs//CIGS Stacked Solar Cells for Space Applications

To improve the practicality of solar cells for outer space, a few practicable advancements are required. Cells with the highest levels of efficiency are characteristically used in space to minimize weight. They must also be engineered with high resistance to the severe conditions of space and particularly high radiation levels. Solar technological research in the past had pointed to InGaP/GaAs/InGaAs cells as a potential cell type to meet these requirements. Kawakita *et al.* (2016) have identified that one issue and area for improvement within this cell type is the InGaAs sub-cell component. Improving this sub-cell type will both improve electrical output and, potentially, reduce radiation resistance.

Copper indium gallium diselenide (CIGS) has been identified as a potential substitute for the InGaAs sub-cell component. CIGS thin film cells have high conversion efficiencies and, relative to other thin fill cell types, extremely high resistances to radiation and past studies have concluded that CIGS cell types show no degradation in space for up to 10 years. The National Institute of Advanced Industrial Science and Technology (AIST) has fabricated InGaP/GaAs/CIGS solar cells with efficiency levels reaching as high as 24%.

Kawakita *et al.* fabricated this cell type and executed radiation tests to examine the difference in radiation resistance between traditional space cell types and the new InGaP/GaAs/CIGS ones. They observed degradation patterns in typical InGaP/GaAs/InGaAs cells and InGaP/GaAs/CIGS cells when the surfaces were exposed to highly energetic electrons. Although degradation was observed in InGaP/GaAs/CIGS cells, it was relatively small in comparison to the degradation observed in InGaP/GaAs/InGaAs cells. This result shows that the change from InGaAs to CIGS improved the radiation tolerance while producing similar efficiencies.

This study is the first to show that CIGS bottom sub-cell type cells have the potential to be extremely effective space cells. These cells have both high levels of efficiency and high levels of radiation resistance. Developing this cell type will help further space cell technology and, currently, provides a new baseline for radiation resistance in space cell types.

Post-annealing of MAPbI3 Perovskite Films with Methylamine for Efficient Perovskite Solar Cells

Perovskite film solar cells have gained recognition due to their excellent optoelectronic properties. Organo-lead-halide perovskite (OHP) based materials have contributed to the success due to their ability to be easily processed plus their desirable mechanical properties. Power conversion efficiency (PCE) of OHP solar cells has reached 22.1% due to recent advances. The most important factor that influences the performance of perovskite cells is the quality of perovskite films. One reason for poor quality is grain boundaries (GBs). GBs form as a result of cracking in the crystalline structure. The GBs make the cell vulnerable to local chemical environments as well increase the amount of surface recombination. For these reasons, GBs negatively affect efficiency and device performance. This problem has been combated in the past by using additives and anti-solvents to help regulate perovskite crystallinity. Instead of focusing on additives, Jiang *et al.* (2016) combat this issue with in a new way: post-annealing.

Jiang and his team utilized methylamine post-annealing (MPA) for the treatment of a specific OHP solar cell. The MPA process proved to be successful for several reasons but none more important than its ability to allow residual solvents to evaporate from the perovskite film during annealing, leading to a higher diffusion rate of methylamine, and thus, a more uniform surface area. Impurities on the perovskite films were greatly reduced using MPA. Continuity between grains was also stimulated using MPA, leading to less notable grain boundaries.

When compared to other annealed films, the lifetime of the MPA-processed films was increased by three times. Efficiency also skyrocketed, increasing by 43.1% when compared to more typical thermally annealed films. The more uniform surface area also limited the number of recombinations on the surface of the film. Recombination resistance increased by over 10 times when compared to more classically annealed films.

This technique may not only be useful for one type of cell either, as Jiang showed that the MPA technique increased efficiency and stability in both planar and meso-structured perovskite solar cells. This new technique can be utilized in all perovskite films, no matter the fabrication method. Overcoming crystalline instability and surface recombinations is crucial to continued solar efficiency growth and utilizing the MPA process is a great first step to do so in perovskite films.

Efficient and Stable Solution-Processed Planar Perovskite Solar Cells via Contact Passivation

Perovskite cells have been extensively researched due to their low fabrication costs and high electrical conversion rates. This cell type offers compatibility with a wide-range of flexible substrates and the potential for tandem devices that are perovskite-based. The highest performing perovskite cells (~22.1% efficiency) rely on high temperatures for the synthesis of electron-selective layers. Unfortunately, high temperatures raise manufacturing costs and limit the potential cell compatibility.

Current research has worked to lower the temperature required for electron-selective layers without lowering efficiency. TiO_2, ZnO and SnO_2 have commonly been used for synthesis of electron selective layers at lower temperatures. Although these cell types are easily fabricated, their long-term operational stability has remained inferior. Tan *et al.* (2016) reasoned that this inferior stability stems from imperfect interfaces and recombination between the illumination side and the perovskite film.

To combat the issues associated with low temperature perovskite solar cells, Tan's research team has devised a simple passivation method, utilizing chlorine-capped TiO_2 films. The chlorine additive is extremely important, as it enhances grain-boundary passivation and substantially reduces interface recombination. When compared to other TiO_2 based processes, the size and quantity of grain boundaries was greatly reduced. The chlorine-capped films also showed no parasitic absorption loss over the entire visible spectral range.

These cells remained very efficient as well, with certified efficiencies of 20.1% and 19.5%. They also stayed efficient for long periods of time. Surprisingly, the most efficient cells fabricated by Tan's group (PCE > 20%) displayed the best operational stability, retaining 90% of their initial performance after 500 continuous hours at maximum power settings.

Passivating perovskite films with chlorine-capped TiO_2 lessens surface recombination and improves binding between films, leading to a more uniform surface area. Tan's cell model improves past research on low-temperature perovskite solar cells and should immediately be applied. The model allows for low manufacturing costs, high efficiency and, most importantly, a high level of stability.

Improved Performance and Reliability of *p-i-n* Perovskite Solar Cells via Doped Metal Oxides

Perovskite cell types have emerged as potential low-cost photovoltaics. In just six years, with increased research, power conversion efficiencies rose from 3.8% in 2009 to over 21% in 2015. In order to maximize efficiency, a homogeneous electron-selective layer is desired. This allows for minimized surface defects and recombination. It also provides a smooth point of contact for the metal electrode located at the top of the cell. For this reason, the choice and fabrication process of the electron-selective contact layers play a massive role in device efficiency, stability, and functionality.

Inorganic layers made up of earth abundant materials have been reported to improve surface homogeneity and device functionality. Metal oxides are commonly used in these thin layers. Unfortunately, metal oxide infused layers have limited electrical conductivity and because of this, only very thin layers (<50 nm) can be used in solar cells.

One way to potentially avoid this conductivity problem is through doping. Doped metal oxides have enhanced electrical conductivity and allow for the fabrication of thicker conductive layers (>150 nm). Doped metals also provide ideal selectivity and large areas that are highly efficient. Savva *et al.* (2016) fabricated perovskite cells using thick, low-temperature aluminum-doped zinc oxide (AZO) and showed that these cells significantly increase device functionality and reproducibility. Cells with AZO exhibited an average PCE of 20% higher than cells without the AZO layer. The high levels of efficiency were maintained over large area devices. Device stability was enhanced when using AZO as well, as initial PCE values were maintained for over 1000 hours of exposure. The research will not stop with AZO layers either, as this is only one potential doped metal combination.

The findings associated with doped metal layers may have wide reaching consequences. The fabrication of AZO layers and other layers like it may allow for the use of more stable metals than aluminum (Al) as the top electrode in perovskite solar cells. If this step is possible, extremely stable *p-i-n* perovskite solar cells may be fabricated sooner than expected.

Incorporation of Rubidium Cations into Perovskite Solar Cells Improves Photovoltaic Performance

The research on low cost perovskite solar cells has steadily increased over this past year. Perovskite cells have now achieved power conversion efficiencies (PCE) of over 22%. Perovskites used in solar cells have a structure that consists of one monovalent cation (A), a divalent metal (M) and an anion (X). The formula comes out to AMX_3. The most efficient perovskites are Pb-based. Increasing perovskite complexity is motivated by the desire to improve stability. Researchers hope that by adding more inorganic elements and increasing entropy of mixing, stability can be increased. Increasing entropy of mixing can help stabilize ordinarily unstable material. For example, the non-active phase of $FAPbI_3$ can be avoided using small amounts of unstable $CsPbI_3$. Combinations of Cs, MA and FA cations have been explored but ineffective. In order to continue progression in stability, different cations need to be explored.

When looking at tolerance factors, one can easily identify $CsPbI_3$, $MAPbI_3$ and $FAPbI_3$ as established perovskites. $RbPbI_3$ has a desirable oxidation stability but has never been used in perovskites because its tolerant factor is slightly different than the established perovskites.

Saliba *et al.* (2016) proposes the use of Rb^+ in perovskite films. The embedding of Rb^+ occurs into the cation cascade to create highly stable material. Using this process and embedded Rb^+, efficiencies of 21.6% were reached. Losses were limited by this process as well. Perovskite cells embedded with Rb^+ had a loss-in-potential of 0.39 V. In comparison, 0.4 V is the average loss-in-potential for standard, commercial silicon cells. What is much more important and pertinent to many of the questions surrounding stability of perovskites is that these Rb^+ embedded cells maintained 95% of their initial performance at 85°C for 500 hours. This occurred under full solar illumination and maximum power settings. The cell also emitted well in the IR/red spectral range, making this cell one of the most efficient LED perovskite cells fabricated.

Solution-Processed Organic Tandem Solar Cells with Power Conversion Efficiencies >12%

The amount of research on organic solar cells (OSC) has risen substantially over the last couple of years. This is due to their many advantageous characteristics such as relatively low cost, flexibility, lightweight, and solution-based fabrication. Past research has focused on material design and device optimization and advancements in these areas have led to OSCs reaching power conversion efficiencies (PCE) between 10 and 11% for single-junction cells. Efficiencies for OSC are lower than for other cell types, due to the limited absorption range of organic materials in the cells. This limited range leads to a large amount of unused solar radiation, which can potentially be increased by fabricating a tandem cell.

Tandem cells are typically constructed using two single cells and a connecting layer. Ideally, tandem cells will be fabricated where there are no potential losses in the connecting layer. Lots of research has been conducted to find the ideal interconnection layer. Every layer tested thus far has certain disadvantages that have limited the PCE's associated with their performance in tandem cells.

Li *et al.* (2016) fabricate a solution-processed layer consisting of ZnO nanoparticles. The layer immediately showed signs of improvement in the OSC cells. The researchers tested layer thicknesses between 90 and 150 nm. Across all thicknesses, the tandem devices showed voltage values of 1.62V. After testing the sub-cells, 1.62V was determined to be almost identical to the sum of the voltages of the two sub-cells. Efficiencies of over 11%, and up to over 12%, were found across layer thicknesses. This shows that the cell performance was independent of layer thickness, which is especially important when considering the potential commercialization of this cell type.

Tests on performance of the tandem devices under different illumination intensities were also conducted. These tests simulated the varying levels of solar illumination cells actually experience. At relatively high light intensities, Li's OSC tandem cells maintain efficiencies of over 11%. When the amount of light intensity is decreased, the same cells reach efficiencies of up to 12.96%. Under varying light intensity and illumination, the tandem cells continue to work well to consistently and efficiently harvest light energy. These advantages using a ZnO-based solution-processed interconnection layer may create a likely alternative in producing an OSC cell for commercialization.

Conclusions

The solar industry has seen a continued growth in the stability, longevity and efficiency of multiple different cell types. Perovskite cells have most notably experienced a rapid rise of efficiency and this should only endure with intensifying research. Many of the breakthroughs from this past year stemmed from areas of investigation new to even the most experienced researchers and highlights the complexity and diversity of issues confronting the industry and a more efficient solar cell. With this said, there is already hope for huge breakthroughs on the near horizon. Recently, researchers from MIT have begun to detail a process where normally emitted heat is recouped by the solar cells and utilized. This process may double solar cell efficiencies. In the coming years, as solar research continues to expand around the globe and in the United States, solar cells will only continue to grow in strength and capability. They will also expand in popularity as an alternative, sustainable and cost-effective energy source.

References Cited

Jiang, Y., Juarez-Perez, E. J., Ge, Q., Wang, S., Leyden, M. R., Ono, L. K., Raga, S. R., Hu, J., and Qi, Y. (2016). Post-annealing of mapbi3 perovskite films with methylamine for efficient perovskite solar cells. Mater. Horiz., 3:548–555. **doi:** 10.1039/C6MH00160B

Jin, G., Sun, Y., Wang, Y., Zhang, D.W., Chen, L., He, C., Boeker, J., Hong, J., Chen, S., Liu, R., Lv, Y., and Chen, J., 2016. Ion Implant Technology for State-of-the-art High Efficiency Solar Cell Applications. 2016 16th International Workshop on Junction Technology (IWJT), 59-63.

Kawakita, S., Imaizumi, M., Makita, K., Nishinaga, J., Sugaya, T., Hajime, S., Sato, S., Ohshima, T., "High efficiency and radiation resistant InGaP/GaAs//CIGS stacked solar cells for space applications," *2016 IEEE 43rd Photovoltaic Specialists Conference (PVSC)*, Portland, OR, 2016. 2574-2577. doi: 10.1109/PVSC.2016.7750113

Li, M., Gao, K., Xiangjian, W., Zhang, Q., Kan, B., Xia, R., Liu, F., Yang, X., Feng, H., Wang, N., Peng, J., Zheng, H., Liang, Z., Yip, H.L., Peng, X., Cao, Y., and Chen, Y. (2016). Solution-Processed Organic Tandem Solar Cells with Power Conversion Efficiencies >12%. Nature Photonics. doi:10.1038/nphoton.2016.240

Saliba, M., Matsui, T., Domanski, K., Seo, J.-Y., Ummadisingu, A., Zakeeruddin, S. M., Correa-Baena, J.-P., Tress, W. R., Abate, A., Hagfeldt, A., and Gratzel, M. (2016). Incorporation of rubidium cations into perovskite solar cells improves photovoltaic performance. Science. doi: 10.1126/science.aah5557

Savva, A., Burgues-Ceballos, I., and Choulis, S. A. (2016). Improved performance and reliability of p-i-n perovskite solar cells via doped metal oxides. Advanced Energy Materials, 6(18):1600285–n/a. 1600285. doi: 10.1002/aenm.201600285

Shinde, O.S., Funde, A.M., Agarwal, M. *et al.* (2016) Emitter passivation of silicon solar cell via organic coating at room temperature. J Mater Sci: Mater Electron 27: 12459. doi:10.1007/s10854-016-5706-8

Tan, H., Jain, A., Voznyy, O., Lan, X., Garcia de Arquer, F. P., Fan, J. Z., Quintero-Bermudez, R., Yuan, M., Zhang, B., Zhao, Y., Fan, F., Li, P., Quan, L. N., Zhao, Y., Lu, Z.-H., Yang, Z., Hoogland, S., and Sargent, E. H. (2017). Efficient and stable solution-processed planar perovskite solar cells via contact passivation. Science. doi: 10.1126/science.aai9081

Section IV—Human Health and Climate Change

Measuring and Speculating on Health Impacts of Climate Change

Alexander Brown-Whalen

Climate Change is one of the most pertinent matters and areas of study in the twenty first century, as its implications are bleak and of serious magnitude. Perhaps the most important inquiry of this subject is how climate change will impact the health and physiological well being of people throughout the world. Unfortunately however, measuring the immediate and direct impacts of climate change on human health is difficult to study. Primarily, the causations of health outcomes on populations are difficult to isolate, particularly considering the complexity of factors impacting vulnerability to negative health outcomes. Furthermore, non-climatic variables such as political and socio-economic factors can often negate potential health outcomes associated with climate change through the use of strong public health resources and social safety nets that ensure healthy living conditions for vulnerable populations. It is for this reason that Industrialized nations are given less attention in this field of research, despite the inequitable dynamic of such nations being the major contributors to climate change globally.

Despite the complexity and importance of non-climatic factors on health impacts, there are numerous—albeit incomplete—ways to study the direct and immediate impacts of climate change on human health outcomes. Some of the approaches explored in this chapter include: 1) incidence rates of infectious disease with respect to localized climatic variables, 2) cross sectional health outcome surveys inclusive of relevant climatic events, 3) health outcomes related to increasing temperatures and UV levels, and 4) aerosol reduction projections with respect to health impacts. One of the major challenges in this research is its reliance on predictive modeling. Predictive modeling is done by analyzing the trends of available historical data, and using this [often mathematical] trend analysis to formulate a future trajectory of such trends. While this is a critical tool in understanding the possible paths of climate change and related health outcomes, it is also speculative and prone to distortion by non-climatic factors. Thus, it is important to have a balance of modeling and non-modeling based research, that encompasses both big picture and immediate frameworks for studying the health impacts of climate change.

Climate Change and Health in Bangladesh: a Baseline Cross-sectional Survey

Health outcomes related to climate change have been understudied in low-income countries, despite the particular importance to such research. Bangladesh, coming in 6th place on the IPCC's long term climate risk index, has had sparse investigation on the subject. However, Kabir *et al.* 2016 used a cross sectional study, sampling 6,720 households within rural areas, to measure the health outcomes associated with recent climate change in Bangladesh.

Multistage cluster sampling was used to recruit participants from seven different vulnerable districts within Bangladesh. There were 223 different villages sampled from, all of which were rural. For each household, a trained interviewer administered a questionnaire designed to collect household data on socio-economic information, general health and living information, personal experiences of extreme weather events, and an array of health variables associated with climate change. These data were subsequently analyzed statistically using STATA 13.

Kabir *et al.* 2016 found that roughly 71% of respondents believed food crop production had decreased over the previous decade, 96% of respondents had experienced extreme weather events in their locality, 45.2% of respondents had been homeless for more than a month in the past ten years due to floods and cyclones, and of those respondents, 40% of them were displaced twice or more. Additionally, rates of diarrhea and pneumonia for children under 5 were particularly high—31.4% and 23.8% respectively. However, this is speculated to be largely related to the level of vulnerability within the communities sampled. Lastly, 295 cases of malaria were found in non-endemic areas, supporting the idea that climate change has been causing the spread of infectious disease to previously non-endemic areas.

The findings of this study suggest that marginalized populations in Bangladesh are particularly vulnerable to climate change-related health impacts, in addition to comparative negative health outcomes generally. Therefore, Kabir *et al.* 2016 suggest serious adaptation by the country's public health system to better support these communities and respond to risk of climate change.

The findings generally suggest that climate change negatively affects human health, although more research should be done on the subject. Overall, Kabir *et al.* 2016 suggest that this study should serve as the basis of a future cohort study, ideally measuring the climate changes in depth to use for health outcome analysis.

Climate Change and Agricultural Workers' Health in Ecuador: Occupational Exposure to UV Radiation and Hot Environments

Despite the magnitude of discussion and research on climate change, inadequate attention has been paid toward the impact of climate change on the health of agricultural workers in non-industrialized countries. Arjona *et al.* 2016 reviewed the contemporary data and research on this subject in Ecuador, concluding that the negative health impacts for agricultural workers will likely increase with climate change. Despite the small size of Ecuador, it possesses diverse climatic regions. Two of these regions—the Ecuadorian coast and the high altitude Andean mountains—are most susceptible to climate changes. The concerning impacts of climate change on these regions differ: increased temperature in the Ecuadorian coast, and increased UV radiation in the Andean mountains.

Average temperatures in Ecuador have been increasing with the global trend, and the Ecuadorian coast is of particular concern. Typically, within this region agricultural workers work in temperatures of 25–35°C with 60–80% humidity. In addition to the working environment, workers often lack sufficient water and electrolytes. Consequently, dehydration is rampant—particularly within the exploitative sugar cane industry—and often leads to a particular syndrome called "revival", requiring immediate medical care. Furthermore, this population experiences a high rate of kidney diseases. Chronic kidney disease—a widespread kidney disease found within populations of agricultural workers throughout Mesoamerica in general—has been studied extensively, and the numerous hypothesized mechanisms of causation derive from dehydration, heat stress, and physical exhaustion. Thus far, the best understood method of prevention is adequate hydration and decreased exposure to heat stress.

Climate change is theorized to be increasing UV radiation levels by hindering the recovery of stratospheric ozone depletion or by altering UV absorbing tropospheric gases, aerosols, and clouds. Three stations were set up—two in Andean mountains and one in the Ecuadorian coast—to measure the Global Solar UV Index (UVI) in 2014/2015 using solar UV radiation sensors. While all three of these regions experienced substantially higher maximum UVI levels than what is considered "extreme" by the World Health Organization, only one of them—Izobama, one of the two Andean region stations—had noticeably higher UVI levels in 2014–2015. However, these data were only taken from limited months within a two-year timespan, and should be recognized for their limitations. In contrast, in Cuenca, Loja, and Quito—UV susceptible Andean regions—skin cancer has increased continually over the past 10 years. This alone should be cause for increased research in the subject, given the high prevalence of skin carcinomas within these regions. Moreover, many studies have shown that this risk is perpetuated for outdoor agricultural workers. Considering the global trend of climate change, the lack of data on occupational health for Ecuadorian agricultural workers, and the health risks associated with working in these susceptible environments, there is a clear need for further investigation and implementation of preventative measures.

Climate Change Influences Potential Distribution of Infected *Aedes aegypti* Co-Occurrence with Dengue Epidemics Risk Areas in Tanzania

Tanzania experienced four dengue outbreaks between 2010 and 2014, with 961 recorded cases for the 2014 outbreak. Of these recorded cases, 99% occurred within the coastal city of Dar es Salaam—the largest city in Tanzania. Mweya *et al.* (2016) used bioclimatic and infected mosquito data from this outbreak to identify risk areas and project future changes in risk from climate change within Tanzania.

From May to June of 2014, adult *Aedes* mosquitoes were trapped using mosquito magnets within Dar es Salaam. Subsequently, 330 pools of 10 mosquitoes were taken from 27 locations and tested for dengue using real time reverse transcription polymerase chain reaction (qRT-PCR). The testing was done for collective pool RNA, detecting infection for respective pools rather than individual mosquitoes. Of the 330 pools, 27 (8.18%) were found to contain dengue.

Data for 19 different bioclimatic variables— representative of 1950–2000—were used for analysis with respect to mosquito collection data. These variables were specified to 1 sq. kilometer for each location of mosquito collection. Ecological Niche Modeling was used to analyze the relationships between bioclimatic variables and infected mosquito incidence. The models were then used to extrapolate unsampled risk areas considering the identified bioclimatic variables. Additionally, the models were used to predict 2020 and 2050 scenarios of dengue using climate change projections from Coupled Model Intercomparison Project Phase 5.

Mweya *et al.* found 12 variables to be indicators for dengue risk areas, which were used for extrapolation of risk. For the current scenario model, precipitation of the driest month and annual temperature change had the greatest impacts on risk. Future scenario models differed slightly, identifying precipitation of driest month and mean diurnal range for greatest impacts on risk. The results indicated that current risk areas were highly localized to coastal regions such as Dar es Salaam and Zanzibar. However, 2020 and 2050 predictions suggested not only an increase in risk intensity, but that risk areas would increase and shift toward central regions of Tanzania. Overall, the findings show that climate change will both perpetuate dengue risk areas and increase the risk intensity of current risk areas. Nonetheless, the authors suggest that their findings be considered carefully, as their models are confined to bioclimatic factors.

Climate Variation Drives Dengue Dynamics

Human dengue incidence has increased by a factor of 30 in the last 50 years. Dengue proliferation is widely understood to be influenced by temperature, rainfall, and socio-economic factors, however, quantitative analysis of these variables and their relationships are sparse. Xu *et al.* (2016) explored these relationships quantitatively in Guangzhou, China—the city with the most dengue cases in China—using data from 2005 to 2015. Days of rainfall and average maximum daily temperature (by month) were used as the independent variables, with dengue presence as the dependent variable. The dependent variable was further separated into an indirect outcome—mosquito population levels (mosquito density)—and a direct outcome—dengue incidence. Mosquito population levels were then also used as an independent variable in the prediction of dengue incidence. Data for rainfall and temperature were collected from the China Meteorological Data Sharing Service System, mosquito density data were collected from Centers for Disease Control in Guangzhou, and dengue incidence data were collected from the China National Notifiable Disease Surveillance System. For data analysis, SEM, GAM, and ZIGAM statistical modeling were used to formulate predictions. Considering that there was a sharp increase in dengue incidence in Guangzhou from 2013 to 2015, data from 2005 to 2012 were used to formulate a statistical model—using rainfall and temperature data to predict dengue incidence and mosquito density—while the data from 2013 to 2015 were used to test the effectiveness of the model. This approach is not only helpful in foreseeing trends in dengue, but can also be used to quantitatively test the impact of temperature and rainfall on dengue incidence mosquito density. ZIGAM modeling in data analysis from 2005–2012 suggested that increases in dengue incidence could be predicted by increased rainfall, increased daily average temperature and increased mosquito density from the respective previous month. However, the same data analysis suggested that mosquito density was predictable only by number of days with rainfall and mosquito density from the previous month. SEM modeling showed that temperature and rainfall had significantly positive effects on both dengue incidence and mosquito density. In contrast, GAM analysis failed to show a significant relationship between temperature and mosquito density. Unsurprisingly, mosquito density had a positive effect on dengue incidence in each model. The findings were further supported by similar results from the application of the statistical analysis on the entire data set (2005 to 2015). Lastly, in comparing outbreak risk and outbreak intensity, temperature has a significantly positive effect on both dengue risk and intensity. Rainfall, while shown to have a significantly positive effect on intensity, had an insignificantly positive effect on risk. Despite finding strong effects of climate variables on dengue incidence, Xu *et al.* (2016) conclude that the impacts of human activity on local ecology may be more significant than climate variables.

Climate Change, Malaria, and Public Health: Accounting for Socioeconomic Contexts in Past Debates and Future Research

The effects of climate change on malaria are heavily debated, despite consensus on the involvement of temperature and rainfall with malaria transmission. Much of this debate regards the dynamics and interactions between climatic and non-climatic (socio-economic development, public health infrastructure, etc.) factors. Considering the prominence of scenario modeling, non-climatic factors have been largely excluded from climate change research—given the complexity of quantifying them. However, the absence of such factors has made climate modeling possess unrecognized socio-economic assumptions.

Malaria research regarding climate change has centered on future-oriented scenario modeling, accompanied by controversy given the speculative nature of modeling. The political importance of this subject leads to technical debates serving as a proxy for political ones, perpetuated by the magnitude of difference in findings. Much of this disagreement can be illustrated by two points: structural vs. mechanistic modeling approaches, and highland malaria research findings.

Modeling for climate change-oriented malaria research—albeit not exclusive to malaria—can be divided into two types—mechanistic and statistical modeling—fundamentally different approaches, leading to dramatically different findings. Mechanistic modeling focuses on identifying the role of climatic variables on malaria transmission, then uses climate projections to predict malaria changes. This inherently abstracts climatic factors from non-climatic factors, and often results in inflated findings. The second approach, statistical modeling, looks at global trends in malaria distribution for assessing the impact of 'climatic constraints' on malaria transmission. Reciprocally, statistical modeling is scrutinized for neglecting the role of socio-economic achievements in negating climate change impacts.

Highland malaria and climate change research has expanded these modeling contradictions. Conflicting findings in such research in East Africa has sparked intense debate, similarly reflecting subjective identification of variables. Vulnerability is a crucial factor in malaria transmission, pertaining to both climatic and non-climatic variables. While there is consensus on climatic and non-climatic involvement in malaria transmission, minimal research has considered the factor of vulnerability. Moreover, GDP is often used in this limited research as a central component of vulnerability. This is problematic considering the framework's dependency on GDP reflecting equitable wealth distribution and resource allocation—which is frequently inaccurate.

In order for research in climate change and malaria to constructively incorporate vulnerability science—as is necessary—five actions should be taken: 1) the interactions between climatic and non-climatic factors on malaria should be systematized when studying past events, 2) a consistent definition of vulnerability should be developed and used, 3) key socio-economic components of vulnerability should be quantified, 4) stakeholder and interdisciplinary researcher involvement should be promoted, and 5) socio-economic scenarios should be better integrated in this research. Overall, for research in this field to avoid misinterpretation, it is paramount that assumptions and variables be thoughtful and explicit.

Managing and Mitigating the Health Risks of Climate Change: Calling for Evidence-Informed Policy and Action

Climate change not only affects ecosystems, but can also affect socio-economic conditions pertaining to human health and sustainable development. Since the relationships between climatic and non-climatic variables are complex and difficult to dissect, modeling of joint effects has been a challenge to developing quantitative understandings of public health risks. However, extensive research has shown serious public health risks associated with climate change, particularly from extreme temperatures and infectious diseases.

Despite variation derived from vulnerability assessment, it has been found that increased extreme temperatures consistently lead to increased mortality from both hot and cold extremes—albeit predominantly from the latter. These findings are somewhat controversial however, because data for certain confounding variables were not available. Therefore, further research is important, as the implications of these findings are quite important for public health. Moreover, Infectious disease—another major component of climate change-oriented public health research—is anticipated to spread into new areas with progressing climate change, with an increase in the length of seasons associated with infectious disease risk. Similar to mortality risk, the projected outcomes are based on vulnerability assessment.

The World Health Organization demonstrated the magnitude of public health impacts in 2015 when reporting that an estimated increase of 250,000 deaths per year between 2030 and 2050 will occur due to climate change. Furthermore, this number may be conservative, due to the sparse data on health impacts of climate change in low-income countries.

Considering such public health risks associated with climate change, Tong *et al.* 2016 make 10 suggestions for policy and research initiatives: 1) a significant worldwide reduction in greenhouse gas emissions, 2) the development of cost effective public health strategies, with particular emphasis for vulnerable communities and regions, 3) an emphasis of health authorities on seasonal changes in mortality patterns, 4) increased initiatives for disaster management, community resilience, health care facilities, and public health infrastructure in order to minimize public health impacts from climate change, 5) vulnerability assessment research should be further developed, 6) increased effort in quantification of climate change impacts on infectious diseases and vaccine development for such diseases, 7) increased co-benefit analyses for policy strategies, 8) the development of a standard and effective approach for scenario forecast modeling in climate change impacts that includes non-climatic factors, 9) the development of a global health monitoring system that incorporates climate change impacts, and 10) investment from governments, funding agencies and NGOs in the aforementioned public health initiatives regarding climate change.

Tong *et al.* conclude that in order to address the current and increasing effects on human health by climate change, it is important that public health professionals and epidemiologists play an active role in developing plans for responding to adverse health effects of climate change. Furthermore, both scientific and policy communities must work together in formulating and implementing public policy to reduce greenhouse gas emissions in addition to responding to inevitable impacts.

Climate and Health Impacts of US Emissions Reductions Consistent with the Global 2°C Target

While the United States has pledged to reduce emissions related to climate change, little has been officially determined in strategy or acted upon thus far. Despite extensive research on the negative consequences associated with surpassing the 2°C limit, little research has gone into studying the diverse positive impacts that could be brought about by emission reductions consistent with the 2°C mark. Shindell *et al.* (2016) explored the potential achievements associated with transportation emission reductions and energy emission reductions in the United States consistent with the 2°C limit.

Shindell *et al.* modeled emission reductions of 2.7% annually (2015–2050) for national emissions in order to derive emissions data for 2030 consistent with the 2°C goal. Moreover, they modeled a 'clean transport' scenario—75% reduction of surface transport emissions—and a 'clean energy scenario'—63% energy sector emission reduction with respect to the baseline. Additionally, they modeled a baseline scenario for all other emissions using the Representative Concentration Pathway 8.5 scenario, controlling for other emissions.

Subsequently, Shindell *et al.* modeled the effects of these reductions for public health impacts and national economic benefits corresponding to the emission reductions. Overall, they found that while both scenarios decrease global radiative forcing, localized radiative forcing varies—generally with respect to local levels of pollution. Furthermore, within the United States, they found that over time the radiative forcing becomes negative for both scenarios; 2024–2035 for the transportation model, and 2069–2089 for the energy model. The earlier part of each range is attributed to aerosol microphysics modeling and the later part to mass-based aerosol modeling.

The largest impacts on public health—premature deaths in particular—were concluded to be a result of reduced ozone and reduced surface particulate matter with a diameter smaller than 2.5 μm (PM2.5), which were both decreased significantly in each model. Therefore, Shindell *et al.* analyzed these data using concentration response functions from epidemiological studies to speculate health benefits attributed to the two emission reduction models. It was found that roughly 175,000 premature deaths would be avoided by 2030 using the clean energy model, with 22,000 avoided premature deaths annually after 2030. The clean transportation model resulted in 120,000 fewer premature deaths by 2030, with 14,000 annually after 2030. These were mainly attributed to decreases in ozone and PM2.5, with PM2.5 decreases having the strongest impact. Furthermore, 89–91% of the PM2.5 reduction benefits were experienced within the United States, while only 26–33% of ozone reduction benefits occurred domestically.

Lastly, they monetized benefits using the Social Cost of Atmospheric Release (SCAR), finding immediate national benefits to be valued at roughly US$250 billion annually—seemingly greater than implementation costs—and long term benefits becoming between 5–10 times larger than implementation costs. While the hypothetical emission reductions of this paper are far from current U.S policy and comparatively ambitious, these are quite feasible and clearly advantageous.

Climate Change and Grassroots Resistance in Bolivia

Bolivia has been critically affected by climate change in recent years, with highlands experiencing severe flooding and lowlands experiencing drought. These disruptions have resulted in poor agricultural yields that are particularly concerning considering the reliance on agriculture for subsistence and the precarious nature of the postcolonial economy. In response to these impactful changes and the history of neoliberal structural adjustment beginning in the 1980s, a movement has formed based around anti-capitalist and 'living well' (Andean indigenous ideology) sentiments to address climate change in Bolivia and around the world.

Bolivia is currently facing serious climate change threats. These threats include the melting of Bolivia's glaciers, the spread and return of illnesses, rural-to-urban migration, deforestation, and agricultural failure. Such impacts are increasing in magnitude and Bolivia is particularly vulnerable, as it is a resource poor and non-industrialized nation. Meanwhile, Bolivia, like much of the Global South, has contributed negligibly to climate change. In response, a movement of resistance has emerged.

In the 1980s, Bolivia was subjected to neoliberal structural adjustment debt to the International Monetary Fund. Such neoliberal structural adjustment can be briefly characterized by reduced government spending, the privatization of natural resources, dollarization, the elimination of protectionism, and reduced power of the state over transnational corporations. Neoliberalization was furthered by additional neoliberal reforms of the 1990s by Bolivian President Gonzalo Sanchez de Lozada, followed by the attempted privatization of water in 2000. These measures, coupled with the majority indigenous and rural-based population within the postcolonial context, led to strong, organized resistance from several forces of society.

The movement emerging out of these neoliberal reforms originated as an alliance between mainly rural indigenous communities and urban, middle class leftist communities. However, it is now characterized as connecting national and international NGOs with grassroots rural and indigenous organizing. Furthermore, with the election of Evo Morales in 2005 (an at least rhetorically anti-capitalist president), Morales and his administration became members of this alliance, albeit conditional. The Morales administration has both championed anti-neoliberal, environmentalist, and anti-privatization policies and rhetoric, while also continuing some unsustainable development projects, which have neglected indigenous communities. This dialectic characterizes a major contradiction of the Morales' administration; playing both the role of supporter and coworker of the movement, while also the role of opposition in certain cases. Furthermore, this administration has had an overarching approach of using the extraction of natural resources such as oil for redistributive programming. This has experienced strong success, such as the cutting of poverty by 20% in 6 years, but also demonstrates an ironic duality of the administration's approach to climate justice.

Despite the conflicting nature of this administration regarding the emerging movement, Morales and his administration have worked with and been responsive to grassroots organizers in Bolivia. Moreover, this administration has been a remarkably strong advocate of the movement's message and values in the international arena. Specifically, Morales has advocated for the 'living well' ideology internationally, and the connection between capitalism, colonialism, and climate change. Further, Morales has been a staunch proponent of the idea that the Global North must take serious responsibility regarding both emissions cuts and assistance to global south countries in becoming more sustainable. The seriousness of the Morales administration internationally (with respect to climate change) is largely due to the strength of the grassroots movement developing over the past 30 years. Because of the alliances and organizational success of this grassroots movement, this movement has been a strong contributor to the building of a radical international movement confronting climate change. This can be seen in the involvement of this movement with the historic New York climate justice march of of 2014. Overall, this movement can serve as a model for organizing against both neoliberalism and for progressive climate change policy; something particularly relevant in the United States, considering our destructive impact on the environment, and our lack of political accountability surrounding climate change.

Conclusions

It is clear that the climate change and human health is an important subject of study, despite its complex nature. The difficulty with analyzing trends and corresponding relationships involved should only prompt further research that expands the methodologies used to understand the subject. Moreover, there should be greater focus on understanding health consequences through long term, nonphysiological variables that correspond to health impacts such as: 1) agricultural projections under climate change, 2) forced migration due to climate change, and 3) extreme weather events perpetrated by climate change. While these variables cannot be used to speculate specific projections of particular health outcomes, they might ultimately have the strongest impact on human health outcomes. It is crucial that the difficulty in obtaining concrete results—in regard to health impacts—from proxy variables like these does not hold back the amount of research studying them. In order to preventatively address climate change-related health impacts, it will be of paramount importance for industrialized nations to begin radically reducing emissions. On the other hand, While several papers discussed in this chapter imply that health impacts from climate change primarily concern lower-income, non-industrialized nations—in regard to reactive ways to address health outcomes—the factor of socio-economic mediation has more to do with resource allocation than resource concentration. This is clear considering how certain industrialized nations—such as the United States—have inadequate resources allocated toward vulnerable populations, leading to comparatively poor health outcomes. Likewise, certain lower-income nations—such as Cuba—with strong social safety nets can experience remarkable health outcomes and natural disaster responses. Therefore, it is important to recognize that the classical notion of 'development' is not a universal basis for inhibiting negative health outcomes related to climate change on a socio-economic basis. Furthermore, the implication of 'development' as the best solution for negating such health impacts ignores the difficulty for post-colonial nations to 'develop' in the context of Bretton Woods structural adjustment programmes, global neoliberalism and twenty first century imperialism. For this reason in addition to the political and corporate resistance to climate change action, mass-based grassroots organizing will be required to confront the health impacts of climate change.

References Cited

Arjona R, Piñeiros J, Ayabaca M, Freire F, 2016. Climate change and agricultural workers' health in Ecuador: occupational exposure to UV radiation and hot environments. Annali dell'Istituto Superiore di Sanità 52: 368–373;

Fabricant, N., Hicks, K., 2016. The Bolivian climate justice movement: mobilizing indigeneity in climate change negotiations. Latin American Perspectives 43, 87–104

Kabir I, Bayzidur R, Smith W, Lusha MAF, Milton AH, 2016. Climate change and health in Bangladesh: a baseline cross-sectional survey. Global Health Action 9;

Mweya CN, Kimera SI, Stanley G, Misinzo G, Mboera LEG, 2016. Climate change influences potential distribution of infected Aedes aegypti co-occurrence with dengue epidemics risk areas in Tanzania. PLOS ONE 11(9): e0162649, 1–13.

Shindell DT, Lee Y, Faluvegi G, 2016. Climate and health impacts of US emissions reductions consistent with 2°C. Nature Climate Change Letters 6: 503–509;

Suk, J.E., 2016. Climate change, malaria, and public health: accounting for socioeconomic contexts in past debates and future research. WIREs Clim Change 7, 551–568.

Tong S, Conalonieri U, Ebi K, Olsen J, 2016. Managing and mitigating the health risks of climate change: calling for evidence-informed policy and action. Environ Health Perspect 124:A176–A179

Xu L, *et al.*, 2016. Climate variation drives dengue dynamics. PNAS 114, 113—118.

Adverse Effects to Human Health Caused by Climate Change

Jasmine Kaur

Climate change has resulted in deterioration of the world, which is evident by the increased incidences of drought, floods, land-slides, sea levels rising, global warming, melting of snowcaps, and loss of coral reefs. Reports by the Intergovernmental Panel on Climate Change (IPCC) provide evidence of warmer atmosphere and oceans, the rise in sea level, and an increase of greenhouse gas concentration (Shrestha *et al.* 2017). These dangers of climate change go beyond the scope of affecting nature and spread over into socioeconomic impacts, agricultural strain, human undernourishment, environmental burden of disease, and need for effective government networks. Management of climate change needs to be built on a strong foundation in order to reduce the burden of its negative impacts on the human population.

Contributors to climate change range from little to big factors and this chapter explores the dynamics of human health intertwined within those factors. Analysis of impacts of human-caused climate change can influence food productions systems, such as agriculture, fisheries, and livestock. The threat to food security is built on the ability to produce nutritious foods and distribute it equally among the populace (Myers *et al.* 2017). Therefore, the impact of climate change has a deeper effect into the core of society's day-to-day life. The rise of climate change leaves a deadly trail for more extreme consequences to follow. It is a butterfly effect where environmental stress continues to worsen no matter which path is chosen.

Risk management of climate change deals with efforts to reduce greenhouse gas (GHG) emissions, adaptations to the processes of deteriorating climate, and implementation of mechanization systems. Globally the hazards of climate change are widespread, but concentrate specifically on areas where heat exposure, socioeconomic standing, and agriculture laborers are the highest in numbers and impact. "Four billion people who live in hot areas," (Kjellstrom *et al.* 2016) face the harsh incidences of occupation health problems which lead to reduced labor production that prompts cases of human undernourishment. The flaw in the current risk management tools for climate change is the lack of means to provide the appropriate help for the areas where mechanization is needed most.

Low economic standing of the hot spot countries adds to the problem, and the government needs to do its part because effective networking can lead to co-benefitting health and climate change mitigation (deOliveira and Doll, 2016). Too much authority in one place will lead to decisions that will not provide equal attention amongst the human population and advance negative climate changes. The solution to the problem is not simple, but the orchestration of head-strong committees and gripping policies can be a start to fixing just one of the many problems that occur due to climate change.

Environmental and Health Benefits from Change of Dietary Guidelines

A quarter of greenhouse gas (GHG) emissions are reported to come from food systems. The main contributions to GHG emissions are red and processed meat, and low consumption of fruits and veggies. These lead to increased health burdens such as chronic and noncommunicable diseases, ultimately stemming from increased body weight and unhealthy diets. Springmann *et al.* (2016) conducted a comparative analysis where they changed people's dietary guidelines and tracked the health and climate change co-benefits.

The health analysis was based on a comparative risk assessment and environmental analysis was linked to GHG emissions of food regionally. Investigation about the effects of diet change on future environmental and health issues for 2050 used a region-specific global health model with four different dietary scenarios. The first diet scenario was called the reference (REF) scenario and served as the baseline for the experiment, while the second was the implementation of healthy global diets (HGD) by keeping a healthy body weight. The third and fourth diets focused on healthy energy intake of lacto-ovo-vegetarian (VGT) and vegan (VGN) plant-based diets, respectively.

The results found a decrease of red meat consumption by 78% in Western high-income countries and 69% in Western middle-income countries, and similar levels were seen in East Asia and Latin America. Movement away from animal-sourced diets led to decreased mortality with highest number of avoided deaths in developing countries. In comparison to the REF dietary scenario deaths per year, HGD scenario avoids 5.1 million deaths, VGT scenario saves 7.3 million lives, and VGN scenario avoids 8.1 million deaths. In addition, GHG emissions are reduced comparable to the REF scenario emissions that increase by 51% from 2005/2007 to 2050. In HGD, VGT, and VGN dietary scenarios all GHG emissions decreased for 2005/2007 and 2050. Specifically, more than three-fourths of the total reductions were associated with developing countries as 72–76% lower GHG emissions. Problems with results were low relative GHG emission increases in relation to regions that previously had extensive undernourishment, therefore an increase of fruits and veggies intake took more agricultural expansion. Future goals would be to further learn how to manipulate the dietary scenarios to be more economically favorable in the developing countries rather than just the developed countries.

Impact of Climate Change on Human Undernourishment

One of the studied impacts of climate change is increased human undernourishment owing to decreased crop production.

However, the economic implications of climate change-produced undernourishment has not been examined. Hasegawa *et al.* (2016) study future risks of health and economic impacts caused by undernourishment due to climate change.

The study is based on the foundation of previous studies and improving those models to calculate more indicators for longer periods of time with less uncertainty in conditions for future projections. The Asia-Pacific Integrated Model/Computable General Equilibrium (AIM/CGE) economic model is implemented into the framework of the experiment which includes the future crop yield aggregation for each region. The study, run from 2005 to 2100, tests 17 regions and 42 industrial classifications that total 10 agricultural sectors. Four scenarios of climate and socioeconomic conditions were created to investigate the negative climate impacts. Specifically, the study combined two socioeconomic conditions of population and gross domestic product (GDP), and three climatic conditions within the four situations. The economic impact is quantified by GDP and welfare loss, and the human health impact is measured by nine diseases related to being underweight as a child. Thus, there are two economic measurements: change in morbidity and mortality due to nine diseases, and change in mortality compared to economic valuation of lost life.

The results are that regardless of socioeconomic conditions, the negative effects of undernourishment due to strong climate change were unavoidable. However, the negative effect was dependent on the region, and future health effects of undernourishment largely would be attributed to socioeconomic conditions rather than climatic conditions. The end consensus was that in regions like South Asia where the wheat crop consumption directly affects the health and nourishment of the people, the yield changes of wheat will impact the area. Additionally, the economic value of lives lost from climate change undernourishment in 2100, was significantly larger at −0.4% to 0.0% of global GDP and −4.0% to 0.0% of regional GDP, in comparison to additional health expenditure climate change undernourishment.

Climate Change Effects on Global Food Security

Myers *et al.* (2017) highlights the anticipated negative effects of climate change on global food security and the potential research that can prevent or reduce these uncertainties.

The rise of global food demand began in the 1950s, and since then the ability to find new agricultural spaces to fulfill food production for larger human populations has become progressively difficult. The increased strain on food productions continues to threaten global undernutrition with shifts of precipitation patterns that will directly impact crop yield performance.

In the best achievable greenhouse gas emissions scenario, the atmospheric CO_2 concentrations would be expected to be 540 ppm by 2100, however with the current high emissions the levels will be close to 940 ppm in 2100. These future scenarios stress the substantial yield losses of maize and wheat due to overheating, specifically in tropical regions, both on crop yield and on labor force. A resulting decline of animal husbandry further contributes to the issues of project food security.

Further, agriculture will be suppressed by changes in ozone pollution hindering cop photosynthesis, increase winter survival of insect pests causing crop production to fall, loss of pollinators, and decrease nutritional protein content in crops. In addition, the reduction of global fish catch potential leads to decline of fish harvests that will cause deficiencies of vitamin B12 and omega-3 fatty acids. The estimation of 790 million people who do not have sufficient daily dietary energy intake causes concern amongst public health workers.

Influences of human-caused climate change guide the system of food production and distribution. Therefore, in order to diminish future consequences on food security and global health, a look at the past and present is essential. The produced quantity and quality of food from agricultural, fishery, and live-stock operations can guide the pathway of secure food security and how to maintain adequate supply of nutritious foods.

Jasmine Kaur

Government Policies Co-Benefit Health and Climate Change Mitigation

De Oliveira and Doll (2016) broaden the analysis of traditional scientific field work done in two Indian cities, Surat and Delhi, to understand the long-term impact government policies have on both local health and climate mitigation.

They look at the difference between a fast onset and slow onset problem using five key features of effectiveness of organization networks: involvement of multiple levels, network design, appropriate governance, building and maintain legitimacy, and stability. The resilience of the responses of the challenges faced in Surat and Delhi were not treated; rather the emphasis was on various involvement in the governance.

In Surat, the Surat Municipal Corporation (SMC), local administration for the city, faced drainage control in the slums, lack of waste collection, and the pneumonic plague in 1994. Amidst this adversity, the SMC increased leadership obligation in specific zones in Surat and assigned zonal commissioners. It led to the decrease of Slide Positive Rate, a proxy for malaria incidence, to 1.1 in 2011, compared to 9.4 in 1994.

In Delhi, during the mid-1990s, transport emissions contributed heavily to the air pollution with increased particulate matter (PM). The PM levels at first decreased by 15% from 2002 to 2007, but rapidly increased by 75% in the next five years. According to the key features analysis, the issue of climate change sustainability in Delhi was the lack of effectiveness in governance. The centralization of Delhi government in comparison to the decentralized SMC prove that having effective networks are important for reducing environmental pollution and improving urban health.

Linking climate change, health outcomes, and urbanization with government networks puts forth the idea of strength of multiple systems under one scope. The case in Delhi depicts a lack of governmental control leading to increased climate change affects such as pollution. The lesson learned is the importance of understanding the effectiveness of networks and the roles they play in malfunctioning urban systems that directly impact climate change and health.

Relationship of Childhood Gastrointestinal Illness, Untreated Groundwater, and Precipitation

The control of municipal surface water, groundwater, and private wells in the United States varies from place to place. In general, the regulations are minimal and do not mandate federal monitoring of water quality. This has led to reports of 4.3–16.4 million annual cases of gastrointestinal illnesses (GI) caused by pathogens found in public drinking water systems. Amongst the reasons for GI pathogens transported to the drinking water is increased runoff from the increased precipitation association with climate change.

An important concern is the increased number of children contracting GI. In order to reduce cases of childhood GI caused by precipitation changes and drinking water contamination, Uejio *et al.* (2017) conducted a study to examine possible drinking-water treatments in areas with problematic precipitation rates. The study was set in northern Wisconsin in five cities that all had the same access to minimally treated groundwater and healthcare. Summer and fall precipitation rates were the focus of the study, which was on children under the age of 5 years. The International Classification of Diseases (ICD) billing codes, which provided records for specific GI infections, were used to assess the total number of GI healthcare visits in the studied region. The Parameter–elevation Regressions on Independent Slopes Model (PRISM) contributed climate change projections and the Special Report on Emissions Scenarios (SRES) gave insight into future societal changes to properly analyze future rates of childhood GI rising or declining. In order to quantify the GI caused by precipitation, a time series analysis was performed to compare the relationship of childhood GI and precipitation rates in 1991–2010, and the anticipated numbers for 2046–2065. Additionally, a statistical model measured rates of GI with attention to different drinking-water treatments and various effects of climate change for the future.

The results found that drinking municipal water does raise GI rates, and installing drinking water treatment can reduce the risks of GI. Rapid and slow treatment of the municipal water systems were factors into what percentage of the population was affected. The hypothetical scenarios showed rapid water treatment installation reduced the rates of populations at risk by 82.5%, but slow water treatment had little effect, decreasing populations at risk by only 4.3%. The most vulnerable are children in low-income communities who have untreated public groundwater. The work done by Uejio et al offers perspective of the existing climate change and municipal drinking water situation, and a resolution to alter the rates of childhood GI in 2046–2065, through additional treatment to rural community groundwater.

Environmental Burden of Diseases Attributable to Climate Change in Nepal

The people of Nepal suffer from droughts and irregular patterns of heavy rainfall, probably exacerbated by climate change. The average temperature in Nepal is predicted to increase by 0.2°C per decade in a period of 30 years which will directly affect the public health of the Nepalese people increasing water-borne (WB), vector-borne (VB), renal-related, and heart-related diseases.

In order to quantify the health impacts due to climate change in Nepal, Shrestha *et al.* (2017) perform a time-series study to assess the environmental burden of diseases (EBD). The study was conducted in 2009–2014, in ten districts of Nepal that covered all three ecological belts: mountain, hill, and terai. To find the total disease burden, equations to measure the attributable fraction (AF) and attributable burden (AB) were performed to compare to the baseline and future climatic scenario in Nepal. The health burdens are assessed using morbidity and mortality numbers in 22 hospitals in the nearby districts and the climate-sensitive variables which include temperature, rainfall, humidity, and wind speed. Six statistical models were used to link hospital admissions to weather-related diseases. In addition, an analysis was taken of extreme weather conditions and

EBD estimates attributable to temperature from 1985–2014, the baseline period, to 2015–2045, the future period.

For Nepal, an average raise of 2°C in temperature will occur over an 80 year period, which will be associated with rise in WB, VB, and renal disease hospitalizations, and WB- and VB-related deaths. Rainfall will effect WB disease hospitalizations to increase, but VB and renal-related diseases to decrease. An opposite association will occur with rise of humidity, causing VB and renal-related disease hospitalizations and mortalities to increase, and WB disease hospitalizations and deaths to decrease. An increase of wind speed will cause rise of WB, VB, and heart disease hospitalizations, but decline of renal disease hospitalization and WB and VB mortality. These results showcase the current vulnerable positions the Nepalese people face and lack of means to fix their climate related health problems.

Rise of Respiratory Hospital Admissions due to Wildfire-Induced Air Pollution

Effects of climate change threaten to further increase the future health risks of respiratory illness, most commonly associated with wildfires. The air pollutants emitted by wildfires can harm community health in proportion to the range of area burned by the fire. Liu *et al.* (2016) quantify the increasing number of respiratory hospital admissions associated with risks of wildfire smoke in a changing climate.

Measurement of the levels of fine particular matter ($PM_{2.5}$) is used to assess air quality during or after wildfires. The predicted $PM_{2.5}$ levels for 2046–2051, in all 561 Western US counties, is linked with an atmospheric chemistry model that uses an A1B scenario to look at future impacts of climate change. Additionally, the term smoke wave is coined to describe days of elevated $PM_{2.5}$ from wildfires. SW_{98} is the definition of smoke wave days whose $PM_{2.5}$ is in the 98th quantile of all wildfire-specific $PM_{2.5}$ through all present days in the 561 counties. The more intense smoke waves fall into the 99.5^{th} quantile wildfire-specific $PM_{2.5}$ levels and is represented as $SW_{99.5}$. There are two future population projections: one based on current elderly population size, the other for the projected elderly population.

The methods used to analyze the change in respiratory hospital admissions due to wildfire risks include present (2004–2009) and future (2046–2051) health effects of wildfire-induced air pollution under $SW_{99.5}$ days because the relative risk of respiratory burden is more definite for $SW_{99.5}$.

The results estimate a total of 2,600 $SW_{99.5}$ county-days during 2046–2051, with climate change anticipated to be the cause of increased smoke wave days in Western US. Each additional $SW_{99.5}$ day is estimated to lead to 178 respiratory admissions. On average, the highest number of daily admitted patients for respiratory illness is Los Angeles County and the present total of respiratory admissions for Western US is 245,926. Also, the elderly population in the Western US is set to increase from 15.7% present day to 20.9% in 2050. If the changes of growing wildfire risks and population increase are not taken more seriously, the potential health consequences of climate change will further deteriorate.

Analysis of Heat Exposure on Health and Socioeconomic Impact

One big factor of climate change that is reducing human performance and work capacity is heat exposure. Currently, in hot areas, 10% of daylight hours are too hot for work to be performed and by 2085, the loss of productivity working hours will have increased to 30-40%. The hot areas are Africa, Asia, Latin America, and at moderate risk are southeast and southwest United States. In these hot atmospheres the heat transfer of the intrabody to the external environment and away from the body is limited. The influence of heat exposure is causing the core body temperature to rise that leads to serious physiological risks. Mainly affected are the cardiovascular system with limitations of blood flow, increased heart rate, and conspicuous sweating. As climate change progresses the incidences of occupational health problems will rise, and labor productivity and work capacity will fall.

The impacts of these losses vary depending on job intensity, environment, and general location. An instrument that is helpful to assess the heat exposure is the WBGT. The WBGT (wet bulb globe temperature) is a measure of temperature, humidity, wind speed, and heat radiation combined to directly assess the heat transfer rate from the body. In an hourly WBGT of 79°F, heavy labor work capacity is reduced, and a WBGT higher than 90°F, means work productivity is extremely demanding. Excessive heat exposure affects the individual's performance and productivity, and further adversely affects the community and economy. Farmers carry out 80% of the physical labor needed to sustain agriculture resources for low- and middle-income countries. There have been suggestions to limit worker's heat exposure by implementing machinery to substitute for human labor. However, the solution offered ignores the lack of financial resources of the developing countries where these problems are arising.

Low-income populations of tropical and subtropical areas are most vulnerable to effects of heat exposure on work activity. Workplace heat for outdoor occupations, which include construction work, open cast mining, transportation, and community services lead to the problems caused by heat exhaustion. Similarly, indoor occupations that lack air-conditioning such as factory and work-shop buildings, face extreme heat exposure on a daily basis in these hot low-income countries. Often heat strain is ignored, as work productivity is seen as more important than the physiological conditions of the workers. The duration of heat exposure may vary health outcomes from a minimum of heat exhaustion kicking in within an hour of working to a maximum of undernutrition and mental stresses showing up years after the labor was performed. Kjellstrom *et al.* (2016) made estimates of the increasing workplace heat on global and regional areas based on climate change using the WBGT index. They estimated a total global gross domestic product (GDP) loss of US$2.1 trillion in 2030 and a second loss of 23% of global GDP in 2100, with low-income tropical regions being the most affected. To improve the community adaptations to the increased heat levels in hot low-income areas, the next step is to expand the field studies on the physiological impacts that climate change has on occupational health issues.

Conclusions

The state of the current management of global human health with climate change is an issue that fluctuates between troublesome and direct danger. A call to action for immediate change is needed for the increasing mortality and morbidity rates due to illnesses caused by heat exposure, particulate matter, and unsafe water. There should be no more lives lost because of socioeconomic backgrounds causing an increased risk of climate change illness. The unhitched government policies failing to provide the appropriate management of systems need to be revised to benefit the population equally. Future scenarios of the climate change and health have predicted rises in physiological strain if the current levels remain constant. People need to believe in the necessity of change and make it happen.

References Cited

deOliveira, J.A.P. and Doll, C.N., 2016. Governance and networks for health co-benefits of climate change mitigation: Lessons from two Indian cities. Environment International, 97, 146–154.

Hasegawa, T., Fujimori, S., Takahashi, K., Yokohata, T. and Masui, T. 2016. Economic implications of climate change impacts on human health through undernourishment. Climatic Change, 136, 189–202.

Kjellstrom, T., Briggs, D., Freyberg, C., Lemke, B., Otto, M., & Hyatt, O. 2016. Heat, human performance, and occupational health: a key issue for the assessment of global climate change impacts. Annual review of public health, 37, 97–112.

Liu, J.C., Mickley, L.J., Sulprizio, M.P., Yue, X., Peng, R.D., Dominici, F. and Bell, M.L., 2016. Future respiratory hospital admissions from wildfire smoke under climate change in the Western US. Environmental Research Letters, 11, 1–6.

Myers, S.S., Smith, M.R., Guth, S., Golden, C.D., Vaitla, B., Mueller, N.D., Dangour, A.D. and Huybers, P., 2017. Climate Change and Global Food Systems: Potential Impacts on

Food Security and Undernutrition. Annual Review of Public Health, 38, 259–278.

Shrestha, S.L., Shrestha, I.L., Shrestha, N. and Joshi, R.D., 2017. Statistical Modeling of Health Effects on Climate-Sensitive Variables and Assessment of Environmental Burden of Diseases Attributable to Climate Change in Nepal. Environmental Modeling & Assessment, 22, 1–14.

Springmann, M., Godfray, H.C.J., Rayner, M. and Scarborough, P. 2016. Analysis and valuation of the health and climate change cobenefits of dietary change. Proceedings of the National Academy of Sciences, 113, 4146–4151.

Uejio, C.K., Christenson, M., Moran, C. and Gorelick, M., 2017. Drinking-water treatment, climate change, and childhood gastrointestinal illness projections for northern Wisconsin (USA) communities drinking untreated groundwater. Hydrogeology Journal, 26, 1–11.

Effects of Climate Change on Human Health

Shaina Van Stryk

In the past 100 years, the global average temperature has increased by about 0.74°C. Climate change is increasing the risk of extreme weather events, which lead to more disasters that have dramatic direct and secondary effects on public health. The industrialization of production in developed and developing countries has contributed to the acceleration and severity of climate change events. The extreme weather has begun to affect human populations around the world, and as proposed in this chapter, the impacts of climate change will only continue to get worse. Anthropogenic influence on global climate change poses a widespread threat to human health as water resource availability, agricultural yields, biodiversity, food security, and public health continues to be greatly compromised due to the increased occurrences of extreme weather events[9]. Some of the secondary effects of climate change analyzed in this chapter include the change of airborne pathogenic bio aerosol concentrations, increased dengue fever transmission risks, heat exposure, respiratory diseases, and noncommunicable diseases. The impacts of climate change on public health, especially in developing countries, will continue to burden and possibly destroy entire health care platforms that are unable to sustain the future increases in temperature induced health complications. Future weather trends established through generated climate scenarios analyzed in this chapter indicate the effects of climate change on human health have the capabilities to eliminate and disrupt entire communities that are not efficiently equipped with the programs needed to sustain the rapid and severe declines in public health.

Climate Change Effects on Airborne Pathogenic Bioaerosol Concentrations

The presence of extreme atmospheric changes due to global warming has raised questions about possible secondary effects on public health. Climate change has been shown to effect meteorological conditions such as wind speed, global radiation, and humidity. The same meteorological conditions influence the concentrations and transmissions of a variety of airborne bacteria and viruses, which may establish the possibility of a relationship between climate change and airborne pathogenic bioaerosol concentrations.

Environmental conditions that influence properties of pathogenic bioaerosols were examined in a recent study done by van Leukan *et al.* (2016), a team mostly from the Netherlands, a region heavily impacted by Q fever in cattle. The study used five climate scenarios for the time periods 2016–2045 and 2036–2065. They compared their findings to a historical time period of 1981–2010 by using atmospheric dispersion models (ADM) to determine the possible effects of global climate change on airborne pathogenic bioaerosol concentrations, such as the bacterium *Coxiella burnetti* associated with Q fever. The results showed that climate change did affect overall "modelled concentrations". More specifically, the bacteria concentrations decreased in four out of the five scenarios, and showed the largest association with wind speed and global radiation. Hourly averaged effects were recorded and showed positive and negative fluctuation ranging from –67 to +639 percent. The study by van Leukan *et al.,* shows that climate change does influence airborne pathogenic bioaerosol concentrations, which may differ depending on environmental and physical properties associated with specific viruses and bacteria.

Health Impacts of Climate Change on Pacific Island Countries

Global warming affects human health on a large scale through multiple pathways that can leave some countries more vulnerable to its effects than others based on geographical locations and socio-political climates. The Pacific Island countries are among the most vulnerable to climate change due to their susceptibility to changing weather patterns, and their limitations to managing and adapting to health impacts.

A study done by McIver *et al.* (2016), assessed the overall health effects of climate change on thirteen Pacific Island countries through a combination of qualitative and quantitative approaches. The study determined the health impacts of climate change in each country by separating assessments into three phases. In the first phase, "sub-regional inception meetings" in Auckland, New Zealand were held. The science of climate change, and previously used sustainability methods in each country were reviewed in the meetings. In the second phase, a mixed methods approach was used to analyze epidemiological data, stakeholder consultations, and Poisson regression models when possible to determine each country's sensitivity and future vulnerability to overall health impacts of climate change. In the third phase, a "likelihood-versus-consequence" matrix was used to establish a ranking and prioritization of each country's climate-sensitive health impacts.

McIver *et al.* found that each of the thirteen Pacific Island countries had prioritized climate-sensitive health risks for each of the three risk categories (direct, indirect, and diffuse effects). These models suggest that the Pacific Island region may be the first to experience extreme climate-related consequences such as physical inactivity, food insecurity, forced migration, and increased disease frequencies. The small population sizes of some Pacific Island countries also leave them vulnerable to climate change effects that could potentially destroy entire sustainable communities of people. The results and findings of this study reveal how global warming can have such large health implications in areas like the Pacific Island region, where contributions to anthropogenic climate change is miniscule.

Influence of Climate Extremes on Ecosystems and Human Health in Southwestern Amazonia

In Amazonia, two major droughts and four floods have occurred in the past ten years. These extreme climate occurrences have increased infrastructural losses, forest fires, and risk of disease. Aragão *et al.* (2016), studied the impact of extreme climate changes on natural systems and human health in the Amazonian state of Acre. The study implemented PULSE-Brazil (Platform for Understanding Long-term Sustainability of Ecosystems and Health), a UK-Brazil joint initiative, as a tool to analyze recent socio-environmental disasters, and evaluate the effects of temperature increases on the incidence of Dengue fever in Acre.

Aragão *et al.* used PULSE-Brazil data sets and scientific-based platforms to assess three different scenarios affected by climate change. The first scenario was the effect of "drought-associated fires" on respiratory diseases in children under five years-old in Acre state. The study performed a linear-regression model for variables such as rainfall (mm), number of active fires, and aerosol optical depth. These variables were each compared against direct age standardized rates (ASR) for hospitalizations. The study also used a "quasi-Poisson" regression model to measure the incidence of respiratory diseases compared with environmental variables. The second scenario looked at the impact of temperature on Dengue fever. Aragão *et al.* used a linear trend regression to analyze trends in minimum temperatures observed from 1950 to 2012 for each month, and compared them with recorded Dengue fever outbreaks in the state of Acre. The last scenario used CMIP5 projections of rainfall and minimum temperatures to examine possible future impacts of climate change on respiratory diseases and Dengue fever.

The study found that respiratory diseases peaked during higher incidence of "drought-associated" forest fires and atmospheric aerosol emissions, recorded cases of Dengue fever were highest during the wet season of the recorded months, and CMIP5 projections show the possibility of increased Dengue fever outbreaks in the future. Overall, the results show that mitigation and climate adaptation options are needed in order to soften the consequences of climate change on human health in areas such as the state of Acre in Amazonia.

Heat Index Trends and Climate Change Implications for Occupational Heat Exposure in Da Nang, Vietnam

Climate change generates extreme weather conditions that leave manual laborers, outdoor workers, and those living in poorly insulated buildings vulnerable to heat stress. Opitz-Stapelton *et al.* (2016) investigated the potential health impacts of day and night-time temperatures in Da Nang, Vietnam. Apparent Temperature heat indices were used to identify trends and propose future changes in night-time "thermal comfort". The variables collected were 2-m minimum and maximum temperatures, relative humidity, and 10-m wind speed. Meteorological records for the time period 1970–2011 were standardized for averages. The study used six general climate models, and generated two scenarios each for the time periods 1970–2005 and 2006–2055 from the CMIP5 data portal. A quantile mapping technique was used to "statistically downscale" the variables and to address their general circulation model biases. The daytime threshold temperatures for the general climate models were set at 32°C for light work, 28°C for heavy labor, and 37°C for the absolute physiological threshold given by the Ministry of Health thermal comfort averages. The night-time threshold temperature was set at 28°C for all models.

Opitz-Stapleton *et al.* found that for the years 1970–2011, there was an average of 246 days per year that was equal to or exceeded the 32°C threshold set for light work, and an average of 264 days per year that was equal to or exceeded the 28°C threshold set for heavy work. There was a median of 51 nights per year that exceeded the 28°C threshold set for night-time ambient temperatures during the years 1970–1999, and a median of 26 nights per year that exceed-

271

ed the 28°C threshold during the years 2000–2011. "Multi-model median projections" of future day and night ambient temperatures do not fall below 43.9°C after the year 2050. These results show that outdoor and indoor workers, as well as those living in poorly insulated buildings in areas such as Da Nang, Vietnam are at larger risks of heat stroke, as well as a variety of heat induced health complications. Environmental and social changes are needed to decrease the effects of heat exposure shown to worsen in the future.

Climate Change and its Impacts on Human Health in Nepal

Nepal is a country vulnerable to large human health impacts due to extreme weather events such as flooding, droughts, and variability in monsoons. Dhital *et al.* (2016) did a study regarding the potential impacts of climate change on human health in Nepal, consisting of an in depth literature search through the data base Pubmed/MEDLINE, used IPCC and WHO reports, national statistics, and annual reports from the Department of Health Services in Nepal as methods for gathering data.

The study found that about twenty percent of the Nepal population is affected by heat-related illnesses such as diarrhea, dysentery, high fever, and typhoid disease. Twenty-four glacial lakes are predicted to cause major flooding surges, which potentially will affect more than 10,000 people directly. Vector borne diseases such as malaria, kala-azar, lymphatic filariasis, Japanese encephalitis, and dengue fever cases have begun and will continue to rise due to flooding and temperature increases. Nepal has 22 high risk districts for malaria, which contain 8.9 million people. The rising temperatures and precipitation patterns will likely affect the production of crops and other stable foods, which leaves the Nepal population at risk of malnutrition or noncommunicable diseases such as diabetes and coronary heart disease. The study concluded that there is a need for more primary-based data collected to better estimate and adapt to the potential impacts of climate change in the regions of Nepal. Developing countries in general are more susceptible to devastating effects of climate change because they, like Nepal, lack stable social and political infrastructures capable of handling large public health and environmental disasters that could potentially be more intense and frequent. More data would also help to identify priority impacts that need urgent strategies for adapting to future climate change events.

Health Effects of Environmental Exposures, Occupational Hazards and Climate Change in Ethiopia

There is growing concern in Ethiopia over the burden of diseases increased by the effects of climate change, and over the lack of research, policy, and implementation strategies in the country needed to address these effects. A study done by Berhane *et al.* (2016), conducted research on the growing threat of "multi-faceted climate change-related health challenges" in the region. A project period of September 2012 to August 2015 was used to complete a situational analysis and needs assessment (SANA) for Ethiopia. The project focused on three main identifiable areas for analysis, which were air pollution and health, occupational health and safety, and climate change and health.

The SANA for Ethiopia was conducted by collecting data and reviewing related literature from databases such as PUBMED and local journals. Key questions and responses from primary sources were put into a GEO Health database for storage and future use. The structured data collection guidelines established by the GEO Health hub were used for primary and secondary data collection. The study selected qualified data collectors from "academia and practicing experts from government and non-government stakeholders". The progress of data collection was managed by weekly meetings in the months of January to March 2013. The study used both primary and secondary sources in the SANA.

The study found that indoor air pollution caused by "biomass fuel use in poorly ventilated households" had a larger contribution to poor respiratory health than outdoor air pollution. Ethiopia has experienced the re-emergence of climate-sensitive diseases, droughts, floods, and agricultural failings, leading to malnutrition and decreased food security in the region. The established SANA for Ethiopia shows that awareness and preparedness by established policy and implementation strategies for future impacts of climate change in the region are relatively low, which makes the country very vulnerable to negative human health effects.

Attributing Human Mortality During Extreme Heat Waves to Anthropogenic Climate Change

A study, by Mitchell *et al.* (2016), used the previously recorded 2003 heat wave to simulate conditional models in order to examine the health impacts of heat exposure caused by anthropogenic climate change across Europe. The study analyzed sub-daily temperatures and dew point temperatures taken from the Met Office Integrated Data Archive System (MIDAS), atmospheric field data, and information collected from airport weather stations near Heathrow, London and Orly, Paris. For their climate simulations, the study used a 25 km resolution model over Europe that was "embedded in a global atmosphere-only model" (HadAM3P). The study also used MOSES2, the UK Met Office's "land-surface scheme", to represent conditions similar to the 2003 heat wave. The study performed two experiments using general circulation models with impact models. The first experiment focused on simulations of the year 2003 that included climate "forcings" in the model. The second experiment focused on simulations of the year 2003 that only included internal and external "forcings" in the model. In order to relate the number of deaths during a heat wave to heat-exposure directly, the study used a relationship that connects a change in apparent temperature with a change in "baseline mortality rate" using data collected from June to August.

The study found that temperature increases have a higher impact on mortality in Paris than in London. Paris had a "seasonal heat-related mortality rate of about 34 per 100,000", and London had a rate of about 4.5 per 100,000. Overall, the study showed that heat-related deaths were attributable to the heat wave conditions in 2003, and concluded that heat-exposure mortality rates are related to anthropogenic climate change. Climate model assessment projections of future extreme climate events are needed to assess the overall impact climate change will have on human health in European populations if conditional trends continue.

An Analysis of the Potential Impact of Climate Change on Dengue Transmission in the Southeastern United States

Dengue fever transmission and distribution are affected by climate change variables such as temperature and precipitation. A study by Butterworth *et al.* (2016), assessed dengue fever transmission risks influenced by projected climate change variables for 23 selected sites in the southeastern United States. The sites were selected based on characteristics such as population centers, completeness of climate datasets, and risk of tropical disease emergence. Butterworth *et al.* analyzed 20 years of observed daily weather station measurements from the Global Historical Climatology Network Database for each site. The study used projected climate change data derived from the "statistical weather generator LARS-WG5" to propel a Dynamic Mosquito Simulation Model (DyMSiM) that was connected to a virus transmission component. The DyMSiM is a meteorologically motivated, "process-based model" that contains epidemiological and entomological components. The model is used to stimulate the development of the dengue virus when influenced by correlated environmental variables. Development rates of "mosquito cohorts" were calculated using changes in temperature, and water availability influenced by precipitation. The LARS generator and the DyMSiM were used for future and present meteorological data that produced total mosquito populations and human dengue cases for each of the 23 sites. The future climate scenarios were developed by calculating projected average monthly changes in temperature and precipitation between the years 2045–2065. The study calculated GCM data for the years, 1961–1990, to be used a baseline.

Butterworth *et al.* found that dengue transmission could be possible throughout the southeastern United States, and had a higher risk of transmission during the summer months. Of the 23 sites analyzed, those located in southern Florida showed the highest likelihood of dengue transmission. The projected scenarios showed that dengue transmission will continue to have the most activity seasonally during the summer months, with the exception of southern Florida where dengue transmission risk could be year round.

Conclusions

These studies show that climate change will continue to negatively affect human health in many populations around the world, thus establishing it as a global concern. The countries most vulnerable to the impacts of climate change on human health lack a sufficient economic and political infrastructure needed to adequately handle the strains those impacts will have on social conditions and health systems. Direct, indirect, and diffuse effects will continue to increase in severity and negatively affect living conditions in every country around the world if more effective sustainability methods are not acknowledged and widely implemented.

References Cited

"Advocacy and Ethics." Community Health Advocacy (n.d.): 159-65. International Federation of Red Cross and Red Crescent Societies, 2010. Web. 2 Apr. 2017.

Aragão, L. E., Marengo, J. A., Cox, P. M., Betts, R. A., Costa, D., Kaye, N., ... & Anderson, L. O. 2016. Assessing the Influence of Climate Extremes on Ecosystems and Human Health in Southwestern Amazon Supported by the PULSE-Brazil Platform. American Journal of Climate Change, 5, 399.

Berhane, K., Kumie, A., & Samet, J. 2016. Health Effects of Environmental Exposures, Occupational Hazards and Climate Change in Ethiopia: Synthesis of Situational Analysis, Needs Assessment and the Way Forward. Ethiopian Journal of Health Development. 30, 50–56.

Butterworth, M. K., Morin, C. W., and Comrie, A. C. 2016. An analysis of the potential impact of climate change on Dengue transmission in the Southeastern United States. Environ Health Perspect. https://ehp.niehs.nih.gov/wp-content/uploads/advpub/2016/10/EHP218.acco.pdf

Dhital, S. R., Koirala, M., Dhungel, S., Mishra, R. K., & Gulis, G. (2016). Climate Change and Its Impacts on Human Health in Nepal. Journal of Health Education Research & Development, 1-4. https://www.esciencecentral.org/journals/climate-change-and-its-impacts-on-human-health-in-nepal-2380-5439-1000174.php?aid=75803

McIver, L., Kim, R., Woodward, A., Hales, S., Spickett, J., Katscherian, D., ... Ebi, K. L. 2016. Health Impacts of Climate Change in Pacific Island Countries: A Regional Assessment of Vulnerabilities and Adaptation Priorities. Environmental Health Perspectives, 124, 1707–1714.

Mitchell, D., Heaviside, C., Vardoulakis, S., Huntingford, C., Masato, G., Guillod, B. P., ... and Allen, M. 2016. Attributing human mortality during extreme heat waves to anthropogenic climate change. Environmental Research Letters, 11, 074006.

Opitz-Stapleton, S., Sabbag, L., Hawley, K., Tran, P., Hoang, L., & Nguyen, P. H. 2016. Heat index trends and climate change implications for occupational heat exposure in Da Nang, Vietnam. Climate Services, 2, 41–51.

van Leuken, J. P. G., Swart, A. N., Droogers, P., van Pul, A., Heederik, D., & Havelaar, A. H. (2016). Climate change effects on airborne pathogenic bioaerosol concentrations: a scenario analysis. Aerobiologia, 32, 607—617.

Impacts of Climate Change on Respiratory Illnesses

Sonia Shenoi

The effect of climate change on human health is critical to study, especially in terms of studying respiratory illnesses. Various factors affect the environment that subsequently play a role in the health of a population. Such factors include flooding, droughts, wildfires, and winter storms. The patterns of these factors have increased in the past decade to an extent where researchers are apprehensive about the upcoming cycles. The climate drivers include more frequent and intense precipitation, more intense hurricane rainfall, longer droughts, sea level rise-related increases during storm surges, and an increasing amount of snowmelt and rain-on-snow events. These factors affect the infrastructure of a population and can have sundry effects on social determinants of health. Respiratory health is affected in a variety of ways. Increases in air pollution, specifically in fine particulate matter, increase the chance for allergy and asthma outbreaks. Higher temperatures also contribute to more frequent respiratory illnesses with the higher pollen count in the air as well as higher levels of pollution. With current changes in climate change, hospitals have recorded the highest levels of exacerbations of chronic respiratory diseases and in some cases, premature death. Dispersion of air pollutants depends on temperature, solar radiation, and precipitation and affect both urban and rural areas. This chapter delves into the correlation between climate change and respiratory effects on a comprehensive level.

Sonia Shenoi

The Effect of Climate Change on Human Health: A Heat Index Analysis

In the paper *Heat, Human Performance, and Occupational Health: A Key Issue for the Assessment of Global Climate Change Impacts* (Kjellstrom *et al.* 2016), the effect of variations of climate are explored in terms of its effect on human performance in society and human health. In focusing on extreme heat, adverse health effects due to exposures to excessive heat index levels have already occurred in many parts of the world. This can be due to heat waves and other individual factors, but is most directly correlated to the severity of climate change as a whole. It is evident that heat stress is both a health and social hazard. The direct health impacts of heat exposure are usually assessed in terms mortality or hospital admissions.

Clinical effects are a consequence of excessive heat exposure but are not the only consequences. Physiological effects also can reduce human performance and work capacity levels. These results and outcomes have been investigated in numerous international reviews of climate change effects but were highlighted in the Human Health chapter of the recent Intergovernmental Panel on Climate Change (IPCC) assessment of climate change impacts. It has been shown that excessive heat exposures affect not only the individuals but also the local communities and economies.

On a more mechanistic level, it is important to take a look at muscular movement and its heat product. As little as 20% of metabolic energy used by muscles contributes to the muscular external work output; the remainder of the energy used is converted to heat given off as waste. Increased core body temperatures have resulted in excessive sweating which in turn causes dehydration and can have several direct and indirect implications for health and well-being.

In order to quantify the health impacts of heat and climate variation, the analysis is based on a heat index. This method accounts for key climate factors that affect both exposures to heat and human physiological mechanisms such as air temperature, humidity, air movement, and heat radiation. These analyses are imperative to understand how climate change has resulted in increased exposures to intense heat in many parts of the world, and how it has affected human health.

Heat exhaustion and reduced human performance are only a few factors that are affected by climate change. Later in this century, many among the four billion people who live in hot areas worldwide will experience significantly reduced work capacity owing to climate change. The social and economic impacts will be considerable, with global gross domestic product (GDP) losses greater than 20% by 2100. Thus, heat performance and human health are correlated with each other and are a critical study of focus.

The Impact of Extreme Heat on Human Health: A Climate and Health Assessment

Alderman *et al.* (2012) discusses how climate change will increase exposure risk of US coastal populations. Due to hurricane intensity and rainfall rates, sea levels will rise and the resulting factor will be an increase in storms. Certain coastal populations will be more vulnerable to health impacts from coastal flooding to at-risk groups such as older adults, pregnant women, and children.

Some regions of the United States have already experienced impacts that are both costly and dangerous. These include the frequency, intensity, or duration of certain extreme events such as flooding related to extreme precipitation, hurricanes and coastal storms, droughts, wildfires, and winter storms and severe thunderstorms. Climate change projections demonstrate that there will be a continuing increase in the occurrence and severity of some extreme events by the end of this century. For each event type, there exists a discussion of populations of concern that have greater vulnerability to adverse health outcomes.

The severity and extent of health effects associated with extreme events depend on the physical aspects and impacts of the extreme events as well as the human, societal, and environmental circumstances at the time and place where events occur. Two important terms include exposure, which is the contact between a person and one or more biological, psychosocial, chemical, or physical stressors, and sensitivity, which is the degree to which people or communities are affected, either adversely or beneficially, by climate change and variability. Extreme environmental effects will also affect and disrupt the essential infrastructure that exists within a population. Disruptions can include reduced functionality such as poor roads that limit travel, or complete loss of infrastructure such as bridges being washed away. Extreme events can also disrupt access to health care services via damage to or loss of transportation infrastructure. Evidently, climate change will alter the frequency, intensity, and geographic distribution of some of these extremes, which has potentially detrimental consequences for exposure to health risks from extreme events.

A Critical Analysis of Air Pollution Health Effects on Respiratory Health

Nemery *et al.* (2016) discusses the historical air pollution studies to get a general overview of the health effects that can be attributed to poor air quality. The authors specifically reviewed the important respiratory effects, the plausible mechanism and population at greater risk.

Studies reporting daily changes in death counts attributable to short-term changes in air pollution are increasing. There are long-term as well as short-term health effects. Some studies reporting long-term exposure to air pollution have inverse and statistically significant reductions in life expectancy.

The respiratory system is the main portal of air pollution entry and consists of biochemically active tissue involving mediators that induce both local and systemic effects after exposure. The lung is the first organ that is affected, even merely hours after exposure. The respiratory system is burdened with allergies, diseases, and chronic sicknesses such as pneumonia or asthma that can be results of air pollution affecting the respiratory tract. Respiratory and cardiovascular health can be affected in two ways: acutely or chronically. In the acute effect, the lungs are overstressed which can lead to hospitalization, reduced lung growth, reduced small airway function, asthma, lung cancer, and other blockages of the air way. The chronic effect includes atherosclerosis as well as chronic bronchitis. Both affect one's heart rate and blood pressure and elicit premature failure of the lungs. Evidently, this is a negative consequence as lung function is imperative for a fully functioning and healthy body.

Vulnerability and susceptibility to the adverse health effects of air pollution could be related either to variation in exposure between individuals and groups, or to the degree to which individuals or groups may respond to a given exposure. Despite the current research, there evidently needs to be more research on the long-term health effects of air pollution in terms of the respiratory system as well as other bodily systems in order to more fully understand how the effect of a change in temperature, climate change, or environmental change may affect human health.

The Impact of Regional Climate Change on Human Health

The World Health Organization estimates that the warming and precipitation trends due to climate change of the past 30 years claim an average of 150,000 lives annually. This is due to the incredibly close connection between climate fluctuations and human diseases ranging from cardiovascular mortality to respiratory illnesses. Climate change also affects levels of malnutrition, infectious disease transmission, and crop failures.

Global average temperatures are projected to increase between 1.4°C and 5.8°C by the end of this century, along with an expected rise in sea level. The number of people at risk from flooding by coastal storm surges is projected to increase from the current 75 million to 200 million in a scenario of mid-range climate changes. This prediction is based on the sea level rising merely 40 cm. In addition, weather cycle extremes such as rainfall and drought are expected to increase with warmer ambient temperatures.

Exposure to both extreme hot and cold weather is associated with increased morbidity and mortality, compared to an intermediate 'comfortable' temperature range. Hot days occurring earlier in the summer season have a larger effect than those occurring later. Climate influences on regional locations are indicative of change as well. Malnutrition remains one of the largest health crises worldwide and according to the WHO, approximately 800 million people are currently undernourished. Droughts and other climate extremes have direct impacts on food crops, and can also influence food supply indirectly by altering the ecology of plant pathogens.

Exposure to hotter temperatures also increases the presence of disease transmission, such as malaria and dengue fever. Temperature has also been found to affect food-borne infectious diseases. Warmer temperatures allow for the disease to persist and replicate longer, thus increasing the number of transmissions.

Climate Change, Air Pollution, and Respiratory Health

Due to climate change and other factors, air pollution patterns are changing in several urbanized areas of the world, with a significant effect on respiratory health. Studies have shown that the heat wave episodes have consistently gotten larger as heat patterns have increased. The process by which weather is monitored is by the emissions of carbon dioxide and other greenhouse gases, following the industrial revolution and other mechanical events that produced increased levels of pollution. An increase in temperature leads to an increased likelihood of floods, especially flash floods, while the occurrence of heat waves and droughts increase the risk of wildfires and desertification.

In urban areas, climate change is likely to influence outdoor air pollution levels because the generation and dispersion of air pollutants, such as ozone and particulate matter, depend on local patterns of temperature, wind, solar radiation, and precipitation. In some regions, air quality is projected to further worsen due to the increased frequency of wildfires that cause the release of gaseous and particulate pollutants in the atmosphere.

The risk of flash floods, wildfires, and desertification is expected to increase due to higher intensity of climate change. These risks affect the infrastructure of the population affected as well. This includes economic factors, where the infrastructure of buildings or roads is torn down due to intensity of acts of nature.

Changing patterns of disease are occurring in response to changing environmental conditions. It is widely recognized that air pollution has a a significant impact on human health, with a great burden on respiratory diseases, particularly asthma, rhinosinusitis, chronic obstructive pulmonary disease (COPD) and respiratory tract infections. Changes in climate change are expected to further aggravate the effect of air pollution on these diseases. The American Thoracic Society has recently expressed concern about the threat posed by air pollution and climate change on respiratory disease. Current and future impacts on respiratory mortality and morbidity deriving from changes in climate as well as from trends in air pollutants are, therefore, a priority for the researchers, respiratory clinicians and policymakers agenda.

Additionally, growing patterns of trees and vegetation are also being impacted by climate change. The dispersal of seeds through wildfires and the response to lack of water directly affects the amount of vegetation that grows as well as the nutrient level it receives. The air quality is also directly correlated between climate and air quality. High temperatures are often associated with dry weather conditions, which significantly contribute to high ozone levels during heat waves. There is no doubt climate change is directly affecting the world we live in.

Climate Change and Air Pollution: Effects on Respiratory Allergy

Major changes involving the atmosphere and climate, including global warming induced by anthropogenic factors, have impact on the biosphere and human environment. Urbanization with its high levels of vehicle emissions, and a westernized lifestyle are linked to the rising frequency of respiratory allergic diseases and bronchial asthma observed over recent decades in most industrialized countries. Global earth temperatures have markedly risen over the last 5 decades due to the increase in greenhouse gas emissions. Global warming from anthropogenic-derived greenhouse gases has consequences that include public health risks.

The effect of heat waves on mortality is well documented. It has been observed a rapid rise in the number of hot days and severe meteorological events, such as the 2003 and 2012 heat waves when temperatures rose above 35°C, have resulted in excess deaths. There is evidence of an increased number of deaths and acute morbidity, especially among respiratory patients due to heat waves. For every single degree increase, the risk of premature death among respiratory patients is up to 6 times higher than in the rest of the population. The increase in respiratory mortality is larger than total or cardiovascular mortality. Admissions are also apparent for respiratory than for cardiovascular disease. Heat and drought conditions contribute to wildfire risks. Smoke emissions can travel hundreds of kilometers downwind of fire areas, exposing people to a complex mixture of fine particles, ozone precursors, and other health-harming compounds.

The effects of climate change on respiratory allergy are still unclear, and studies addressing this topic are lacking. However, data has been provided showing a projection of plant and fungal reproduction increase that could affect the sensitivity of allergy-related illnesses and trigger them. Climate change is correlated with pollen allergy in several ways. First of all, the rapid growth of plants due to increased temperatures allows more pollen to exist in the air due to an increase in production. This, in turn, leads to an increase in the amount of pollen that is produced by each plant. There is an increase in the amount of allergenic proteins contained in pollen due to the increase in production. Additionally, climate change correlates to pollen allergy due to the increase in the start time of plant growth and therefore the start of pollen production and earlier and longer pollen seasons. These greatly affected the respiratory and allergy illnesses that affect several populations.

Similarly, an increasing body of evidence has demonstrated asthma epidemics are related to thunderstorms. It shows that the occurrence of severe asthma epidemics during thunderstorms in the pollen season, several epidemics of asthma have been reported following thunderstorms in various geological zones, most prevalently in Europe and Australia. Asthma epidemics related to thunderstorms are limited to seasons where there are high atmospheric concentrations of airborne allergenic pollens. Much remains to be discovered about the relationship between asthma attacks and thunderstorms, but there is reasonable evidence in favor of a causal link between them in patients suffering from pollen allergy.

Strategies to reduce climate change and air pollution are political in nature, but citizens, especially health care professionals, must be vocal in the decision process to give solid support for clean policies on both national and international levels. These efforts are crucial for reducing future impacts.

A Closer Look at Particulate Air Pollution from Wildfires in the Western US

Wildfires can impose a direct threat to human health under climate change. Identifying communities that will be affected as well as analyzing the patterns of wildfires will inform the development of fire management strategies, inform the general public, and better prepare disaster programs. Climate change has increased the frequency, intensity, and spread of wildfires. Smoke from the wildfires contains an exorbitant amount of fine airborne particulate matter. There are two types of harm to human health: chronic exposure in which life expectancy can be decreased and the risk of chronic diseases increased, as well as acute exposure which includes negative respiratory symptoms.

The goal of this study was to demonstrate how wildfires can impose a direct impact on human health under climate change and to identify communities that would be most effected in order to develop fire management strategies and disaster preparedness programs. The term $PM_{2.5}$ stands for particulate matter, or fine matter that exists within the air. The term "smoke wave" stands for a period of two consecutive days with high levels of wildfire-specific particulate matter. This was the first study to look at the daily particulate matter in terms of climate change. First calculated the future change in area burned for each model separately and then determined the median changes for the model. Then calculated both non-wildfire and wild-fire-specific PM. The authors used estimates of biomass burned derived from median area burned together with emission factors for carbon species.

Wildfires are estimated to contribute to roughly 18% of the atmospheric emissions in the US. Estimating the levels of air pollution can be challenging, even with present-day technologies. The methods in which this was executed included making a model in which there was a fire-prediction. Current and future estimations were made using the simulated meteorological fields from the model. The climate models show a large range in their projections of key variables associated with weather conditions conducive to wildfires by the mid-century. To estimate the size of populations for children and the elderly in each county in the future, the US Census survey estimates were combined. An interactive map was created to visualize county-level smoke wave characteristics.

The figures show the number of smoke waves in each county over 6-year periods in the present day and in the future under climate change. The average smoke wave intensity is expected to increase an average of 30.8% and the length of the smoke wave season is estimated to increase by an average of 15 days. More than 60% of counties are anticipated to face more smoke wave days under climate change. The study demonstrated that smoke waves are likely to be longer, more intense, and more frequent under climate change, which raise economic, ecological, and health concerns.

Other results include the discussion that wildfire-specific particulate matter can impose economic burdens by impacting medical care, tourism, property values, and costs of forest suppression. Furthermore, wildfires can cause ecological damage. It can affect the firefighters and fire departments as well: During 2000 – 2002, US federal agencies spent over a billion dollars trying to suppress wildfires. Findings suggest that fire suppression may be needed in 2050s in order to lower air pollution to reduce potential health concern. Evidently, smoke waves are bad for health due to exposure to high levels of particulate matter. Most directly, it will affect vulnerable populations especially children during developmental stages. As aforementioned, acute and chronic exposure make it a real risk

Some limitations to the study include the fact that they did not incorporate the possibility that fire suppression might lead to an unnatural accumulation of forests. Changes in vegetation lead to changes in CO_2 levels, which also directly affect wildfire outcomes and air pollution. Evidently, the change in wildfires as predicted by this model and study propose a serious threat to our future climate.

Documenting Human Health Impacts of Climate Change in Tropical and Subtropical Regions

Climate change is harming human health, and the magnitude of the harm is increasing. This is particularly true in tropical and subtropical regions that are more vulnerable to heightened intensity levels. Nearly all countries situated in the geographic tropics are poor, and therefore have fewer resources to adapt to change in climate. Furthermore, industrial and vehicle emissions in hot, humid cities contribute to poor air quality due to smog, resulting in increased morbidity and mortality from respiratory diseases. Finally, heat, drought, and extreme weather events impact agricultural production and threaten food security, which affects multiple populations that rely on subsistence farming.

In a striking example of the observed effects of climate change, recent research has demonstrated that the Solomon Islands are disappearing due to shoreline change as a result of global sea-level rise, destroying villages and leading to community delocalization. The authors analyze this loss of infrastructure and examine the challenges it presents to multiple health populations. The study uses a variety of surveillance sources to identify multiple direct impacts of this extreme weather event on public health: acute morbidity and mortality. The results demonstrated that mortality risks increased with climate change impacts such as extreme weather conditions, flooding, storm watches, and sea-level rises. The events in the Solomon Islands are not expected to be an isolated occurrence, and climate change will continue to cause epidemiological changes throughout the tropics. Physicians in tropical and subtropical regions including but not limited to Ecuador, Brazil, Argentina, Chile, Peru, Turkey, Cyprus, Georgia, Nigeria, and Indonesia have observed patient health outcomes they attribute to climate change.

The field of tropical medicine will confront some of the first, most widespread, and most pronounced human health impacts of climate change due to the massive effects of a slight increase in temperature. Documenting these health outcomes in localized contexts will enable practitioners to target treatment according to health vulnerabilities, highlight regional adaptation needs for public health agencies, and establish evidence of the earliest and mounting health consequences of climate change for policymakers and the public. More broadly, these efforts are crucial to global health interests and serve as a harbinger of the integral role that the health sector will occupy in climate change adaptation and migration.

Conclusions

Evidently, climate change plays a significant role in the development of respiratory illnesses and in studying respiratory health. The American Thoracic Society and European Respiratory Society have expressed concern about the threat posed by air pollution and climate change within the next 50 years. They have set up warnings that have stirred this field of study. Some of the main causes of respiratory health as affected by climate change include higher temperatures, heat waves and cold waves, droughts, and winter storms. Both combined and separately, these have increased the number of respiratory illnesses, leading to a focal point of study within the significant correlation between climate change and human health.

References Cited

Alderman, K., L. R. Turner, and S. Tong, 2012: Floods and human health: A systematic review. Environment International, 47, 37–47.

D'Amato G, Holgate ST, Pawankar R, Ledford DK, Cecchi L, Al-Ahmad M, *et al.*, 2016. Meteorological conditions, climate change, new emerging factors, and asthma andrelated allergic disorders. A statement of the World Allergy Organization. World Allergy Organ J.

Gualdi S, Navarra A. Scenari climatici nel bacino mediterraneo, 2016. 19–30.

Kjellstrom, T. Briggs, D. Freyberg, C. Lemke, B. 2016. Heat, Human Performance, and Occupational Health: A Key Issue for the Assessment of Global Climate Change Impacts. Annual Report of Public Health. Vol. 37, Pgs. 1–412.

Kreslake, Jennifer; Sarfaty Mona; Maibach Edward, 2016. Documenting the Human Health Impacts of Climate Change in Tropical and Subtropical Regions. The American Society of Tropical Medicine and Hygiene.

Liu JC, Mickley LJ, Sulprizio MP, Yue X, Dominici F, Bell ML, 2016. Exposure to wildfire-specific fine particulate matter and risk of Hospital Admissions in urban and rural Counties in the Western US 2004–2009 Epidemiology.

Nemery B, Hoet PH, Nemmar A. The Meuse Val-ley fog of 1930: an air pollution disaster. The Lancet. 2016;357(9257):704-8.

Patz, J. A. , Epstein, P. R. , Burke, T. A. & Balbus, J. M., 2016. Global climate change and emerging infectious diseases. J. Am. Med. Assoc. 275, 217–223.

Effects of Air Pollutants on Neurological Systems

Thy Annie Nguyen

Ambient air pollutant exposure is one of the primary sources of respiratory and cardiovascular illnesses, but the neurological effects have yet to be studied in detail. This chapter seeks to highlight the novel inclusion of neurological disorders and diseases in the adverse impacts of air pollution. As urbanization increases pollutant production, and climate change decreases atmospheric permeability, long-term exposure to polluted air can result in significant neurological impacts as it penetrates deeper into the central nervous system. These pollutants are believed to inflame areas of the brain and even penetrate the blood-brain barrier, infecting sensitive neurons by inducing neurotoxicity and oxidative stress. The effects of air pollutants on neurological systems are global, and can impact entire populations throughout the world. Despite geography or age, the impact of toxic particulates is severe and causes great concern. From neurodegenerative diseases to neurodevelopment, pollutant exposure generally causes inflammation and adverse neuronal health. Progressive diseases such as Alzhemier's Disease, Parkinson's Disease, and multiple sclerosis have been found in growing incidence in highly polluted areas, and are correlated with increased particulate matter exposure. Inhibition of neuron growth and health has been documented in fetal stages and the age at which full brain development reached is in the early to mid-twenties. This timeframe also allows for a large span in which neuronal growth and health may be compromised by pollutant-induced toxicity. Recent studies have suggested that infant health and cognitive ability are also impacted by areas of high pollutant densities, with children displaying low birth weights and latency in neurobehavioral tasks. Understanding how air pollution is affecting populations at the neurological level is necessary to understand human health in a modern, industrialized world with potential long-term effects.

Pathophysiological Effects of Particulate Matter Air Pollution on the Central Nervous System

Human activities have severely impacted air quality. Whereas cardiovascular disease and respiratory conditions have been the main concern in assessing the adverse effects of air pollution, neurological impacts are recently being studied and are equally important. Wright and Ding (2016) review the many adverse effects of particulate matter (PM) exposure and the implications of increased air pollution in highly urbanized communities. From short-term insults to long-term diseases, it was found that PM from air pollution was a direct cause of a myriad neurological diseases as well as increasing mortality and cardiovascular morbidity. PM air pollution consists of "metals, dust, various organic compounds, and microorganisms suspended within aerosolized droplets" (Ding). When inhaled, ultrafine PM is small enough to cross blood and mucous barriers, travelling to the brain and other parts of the central nervous system. As ultrafine PM enters these sites, it may directly cause insults on neurological systems through inflammation, inhibiting blood flow, decreased brain volume, increases oxidative stress, and much more. For example, long-term exposure to ultrafine PM has been shown to lead to vascular conditions such as hypertension and atherosclerosis. In addition, blockages in blood vessel in the brain have been strongly associated with increased likelihoods of stroke. The World Health Organization suggests that one of the most preventable causes of stroke may be reducing PM exposure, decreasing human mortality by almost 750,000 deaths per year. Looking at long-term neurological diseases, constant exposure to PM leads to an immunological response that stimulates the production of pro-inflammatory cytokines. As a result of constant inflammation, cell proliferation is reduced and toxicity and cell death is more likely. These factors may contribute to diseases such as Alzheimer's Disease, Parkinson's Disease, and other conditions that arise from decreased brain volume and obstructive cell debris and plaques.

Although the exact mechanisms of neurological disease are still not fully understood, there is strong evidence reflecting a relationship between environmental conditions and the onset of disease. Especially in communities with high air pollution and low air quality, aggregation of PM may lead to disproportionately higher incidences of detrimental health effects. Considering the increase of urban development and air pollution, it is necessary to observe how human health may be impacted by climate change and recognize the relationship between environment and health.

Multiple Sclerosis and Air Pollution Exposure: Mechanisms toward Brain Autoimmunity

While there has been significant evidence suggesting the correlation of air pollution and ambient particulate matter exposure with neurological diseases there has been little understanding about the pathology and mechanism of action. Multiple Sclerosis (MS) is an inflammatory autoimmune disease that attacks the central nervous system, but not much is known about the cause and predispositions for disease. However, strong relationships between MS onset and air pollutants including PM_{10}, SO_2, NO_2, and NO have been found. A decrease in air quality has been found to induce inflammation and aggravate existing MS conditions. Mousavi *et al.* (2017) seek to explain these observations by hypothesizing the mechanisms that results in increased MS incidence and progression. The primary cause of MS may lie in inflammation and oxidative stress outside of the brain along the blood brain barrier as a result of outer pulmonary and cardiovascular infection by air pollutant exposure. Normally, the immune response uses oxygen or nitrogen radicals to neutralize toxic cells or pathogens. However, when the immune response is prolonged or dysfunctional, free radicals can escape and destroy other cells, inducing oxidative stress on healthy cells. Microglia that act as the primary immune response in the brain and central nervous system, become activated by the chronic inflammation, leading to deeper, more dangerous effects in the central nervous system. This overactivated immune response ultimately leads to autoimmune disease, with the microglia or other recruited antibodies affecting the myelin sheaths of neurons and causing the symptoms of MS. Especially combined with other factors of air pollution, such as vitamin D deficiency by reducing the availability of UVB, MS can be significantly intensified or induced. The overall effects observed by these mechanisms results in decreased immunological self-tolerance, a disturbance in the innate and adaptive immune response, and the production of antibodies against neurons. Research that supports this evidence includes increases in inflammatory cytokines in the cerebral spinal fluid of children and decreases in antioxidant gene expression in MS patients living in high pollution areas.

Long-term $PM_{2.5}$ Exposure and Neurological Hospital Admissions in the Northeastern United States

Many prior studies have suggested that particulate matter (PM) exposure may induce an inflammatory response that leads to neurodegenerative diseases and cognitive decline. PM has been known to carry heavy metals, induce free radicals, and contain carcinogens. Especially with $PM_{2.5}$ (particulate matter ≤ 2.5 μm) being small enough to potentially cross the blood brain barrier, it is worth studying how these aerial pollutants may affect neurological health. In a study from 1999 to 2010 in the northeastern United States, Kioumourtzoglou $et\ al.$ (2016) found a correlation between long-term exposure in dense, urban cities that produced large quantities of $PM_{2.5}$ and accelerated disease progression in Alzheimer's Disease (AD), Parkinson's Disease (PD), and dementia. Approximately 9.8 million residents in 50 cities were surveyed while air pollution data was collected from the EPA's Air Quality System Database. Although the design of the study prohibited an analysis of the role of $PM_{2.5}$ in disease onset, researchers were able to measure the effects $PM_{2.5}$ concentrations had on the current population of neurological patients who had already exhibited the onset of disease. Results from the long-term survey found that higher concentrations of $PM_{2.5}$ contributed to a significant severity in neurodegeneration in existing AD, PD, and dementia patients. Hazard risk (HR) increased for all three groups with PD admissions HR = 1.08, AD HR = 1.15, and dementia HR = 1.08. In addition, these incidences of disease are more likely to affect an aging population, which tends to be susceptible to environmental insults and is the group most likely to see the onset of neurological disease. $PM_{2.5}$ and other airborne pollutants lead to oxidative stress and inflammation that resulted in aggravated neurodegeneration. With aberrant neuronal death, the onset of diseases such as AD and PD may be more likely and existing symptoms more severe. Kiou-mourtzoglou *et al.* contribute to the current discussion of environmental toxicology with their study assessing how long-term exposure to $PM_{2.5}$ may affect human health and the progression of common neurological diseases. Looking at current populations of humans in relevant conditions, fluctuations of $PM_{2.5}$ were shown to significantly impact mortality and severity of disease.

Traffic-Related Air Pollution and Parkinson's Disease in Denmark: A Case-Control Study

As industries emerge, air pollution also becomes a major adverse health effect, with traffic being one of the main sources of pollutants. Many recent research studies have confirmed a negative effect of air pollution on neurological health, and therefore a long-term study is necessary to understand the growing effects of urbanization and neurological health. Ritz *et al.* (2016) did a case-control study to examine the correlation between location of residency and the incidence of Parkinson's Disease (PD). PD is a movement disorder characterized by tremors, bradykinesia, rigidity, and uncoordinated or asymmetric movement. Its onset is attributed to the loss of dopaminergic neurons in the substantia nigra in the midbrain. Rather than looking at particulate matter, this study focused on traffic-generated nitrogen dioxide (NO_2), nitrous oxides (NO_x), and carbon monoxide (CO). Each of these neurotoxins is mostly correlated with vehicle emissions and street traffic which, because of its proximity to the human population, may be the most significant source of neurotoxic pollutants. Utilizing air pollution measures from 1971 to the date of hospital admission in the residences of each PD patient, a dispersion model was created to correlate the traffic-related pollution in the area to the onset or risk of PD. Results found that in urban cities, there was a high correlation of NO_2 to NO_x and CO, all of which are traffic pollution mixtures and are highly toxic in high concentrations. NO_2 exposures at the highest level of the top 95th percentile suggested a strong correlation to PD (Odds Ratio = 1.92; 95% CI: 1.32, 2.80). Overall, there was a 9% higher risk (95% CI: 3, 16.0%) per interquartile range increase (2.97 µg/m3) of ambient air pollution as seen by NO_2.

Fetal Growth and Air Pollution—A Study on Ultrasound and Birth Measures

While associations have been made between air pollution and cognitive decline and neurodegeneration at later stages in life, some studies have shown that early exposure to air pollutants via particulate matter can cause deficits in learning and memory neurodevelopment in children. Neural network development *in utero* is a necessary developmental stage that ensures the optimization of connected cells and brain function later in life, since neural pruning mostly happens at these early developmental stages. However, when inflammation from environmental toxins is introduced, disturbances in neurodevelopment may arise and lead to long-term deficits. In a longitudinal study by Malmqvist *et al.* (2016), correlations between air pollution and birth outcomes such as birth weight, preterm births, and other gestational complications were assessed. From 1999 to 2009, over 48,000 births in southern Sweden were documented for this study. Using post-natal measurements of head circumference and birth weight, Malmqvist *et al.* were able to compare this study population to the expected birth weights of prior years and the national population. Including ultrasound measurements in this study, a closer look at the full rates of development *in utero* also led to a better understanding of neurodevelopment and air pollutants. Air pollutants were obtained from regional Emissions Databases, and pollutants including NO_x (a neurotoxic material in high quantities due to its induction of oxidative stress, inflammation, and association with cell apoptosis) were used as a measure. Results from this study found a negative association of NO_x and all the ultrasound measures of biparietal diameter, abdominal diameter, and femur length. Both abdominal diameter and femur length were significantly decreased with increased NO_x exposure at –0.10 (–0.17, –0.03) and –0.13 (–0.17, –0.01) mm, respectively, per 10 µg/m^3 increment of NO_x. In measures of head circumference at birth, the negative effects for NOx remained statistically significant at –0.5 mm (95% CI; –0.7, –0.2) per 10 µg/m^3 increment in NO_x. Finally, birth weight was reduced, showing each 10 µg/m^3 increment of NO_x to correspond to an approximately 9 g reduction in birth weight. Overall, it was found that NO_x consistently led to a variety of negative health outcomes in fetal growth during late pregnancies. This suggests that environmental neurotoxicology can affect health outcomes even at fetal stages, and that there may be implications of later deficits due to early health impacts.

Recent Versus Chronic Exposure to Particulate Matter in Association with Neurobehavioral Performance in a Study of Primary Schoolchildren

Ambient particulate matter is primarily studied in relation to neurodegenerative diseases, but its impacts are rarely studied in children. In order to study long-term neurodevelopment, Saenen *et al.* (2016) surveyed 310 children in primary schools in Belgium and potential sources of particulate matter ($PM_{2.5}$ and PM_{10}) in their residences including elemental carbon, organic mass, sea salt, ammonium, nitrate, ammonium sulfate, and mineral dust. Particulate matter was sampled by portable AEROCET devices for recent and chronic $PM_{2.5}$ exposures. Additionally, traffic noise was measured to test for effects of noise pollution and other potential impacts urban traffic may have. All children participated in neurobehavioral examinations at least once, with 89% of the children taking a maximum of three examinations. The tests included the Stroop Test (testing mental flexibility by selecting the correct color name while ignoring the color of the printed name), the Continuous Performance Test (testing sustained attention domain), the Digit Span Test (for short-term memory), the Digit-Symbol Test (for visual information processing speed), and the Pattern Comparison Test (also for visual processing speed). As a parameter for performance, reaction time and total latency were used. Comparing the results from the $PM_{2.5}$ and PM_{10} exposure with neurobehavioral performance, it was found that the only significant associations were between the Digit-Symbol Test and recent inside classroom PM exposure. The total latency increase was 2.05 seconds for $PM_{2.5}$ and 1.9 seconds for PM_{10}. Additionally, there were correlations between decreased performance on the Pattern Recognition Test and recent PM exposure. For chronic PM exposure, it was found that tests for attention domain (Continuous Performance and Stroop Tests) had non-significant adverse associations. The implications of Saenen *et al.* are significant to understand how might PM exposure affect not only adults, but also children. It was concluded that there were adverse effects of PM on selective attention and visual information processing speed (recent PM) and attention domain (chronic PM) in schoolchildren. Overall, this suggests that further study on the severity and impact on neurodevelopment is necessary and that PM exposure not only affects adults, but also children early in life.

Nanoscale Particulate Matter from Urban Traffic Rapidly Induces Oxidative Stress and Inflammation in Brain

Nanosized particulate matter (nPM) derived from combustion engines may be more neurotoxic than larger sizes of PM. In inhalation studies, it has been shown that nPM at a size of less than 200 nm has been shown to physically cross the blood brain barrier in order to aggregate in the olfactory bulb. The concentration of nPM in the olfactory bulb may ultimately lead to oxidative stress and inflammation, which can potentially be a gateway to neurodegeneration in the central nervous system. Cheng *et al.* (2016) experimentally studied the effect of nPM on mice to see how inhaled particulate matter, which replicates urban traffic settings, may adversely affect neurological health. The mice were exposed to aerosolized nPM at increments of 5 hours/day for 3 days/week with total exposure times of 5, 20, or 45 hours total. After various levels of exposure, the brain regions and cell types were assessed with chemical assays. Results showed a rapid increase in oxidative stress in early markers of oxidative stress (4-HNE and 3-NT) between a 5-hour and 18-hour period. In addition, the number of Iba1-positive macrophages was increased by 30% in the olfactory epithelium (OE). nPM also doubled the number of amoeboid microglia, which is associated with activation of an immune response. In qualitative measures, OE thickness and density also decreased compared to the control, suggesting some form of degeneration. The OE showed a 30% increase in 4-HNE that persisted from the 5 to 45 hours of total nPM exposure. In addition, 3-NT showed a trend for increase at 5 hours and was significantly increased by 75% at 45 hours. As an overall trend, markers of inflammation (TNF-a) and oxidative stress increased with increased exposure time. This study supports the hypothesis that nPM may induce neurological damage, but is also significant in that the duration of exposure can help with prevention of mitigation of harm.

Particulate Air Pollutants, APOE Alleles, and Their Contributions to Cognitive Impairment and Amyloidogenesis

Although late-onset Alzheimer's Disease (AD) is the most common form of the disease, its initiation and progression are poorly understood compared to genetically inherited early-onset AD. It is suggested that environmental toxins may play a large role in AD, but the exact mechanism is still unknown. Cacciottolo *et al.* (2017) studied the connection between the genetic factors influencing AD and potential environmental insults that alter the current genetic expression. Apolipoprotein E (APOE) is a protein that has been previously suggested to play a role in AD, because it transports cholesterol to the brain and supports neural structure. In this study, a human epidemiological study is compared to an experimental study with transgenic mouse models for APOE alleles. For the epidemiological study, 3647 women were assessed for cognitive and neurological impairments from 1999 to 2010 along with nanosized particulate matter (nPM at $PM_{2.5}$) levels at their residences. It was found that residents living areas with high $PM_{2.5}$ levels had increased risks of cognitive decline, and those with the $\epsilon 4/4$ APOE allele had the highest risk for global cognitive decline and all-cause dementia. In the mouse studies, varied APOE alleles were tested, which included $\epsilon 3$ and $\epsilon 4$ alleles of the APOE gene. These mice were then exposed to PM or control conditions to test for the relationship of various genes and the effects of environmental pollutants. It was found that the E4FAD mice (carrying the $\epsilon 4$ allele) were the most sensitive to particulate matter, with higher Aß plaque load and plaque densities significantly increasing in comparison to the control and all E3FAD mice. Finally, it was found that chronic $PM_{2.5}$ exposure resulted in significant decreases in glutamate receptors (GluR1) in the hippocampus across all alleles, and especially in the E4FAD mice, which is significant because it shows decreased synaptic ability and function in the primary memory center of the brain. In conclusion, both the human epidemiological and experimental mice studies showed similar results in the relationship between genetics and environment as they can both affect the onset of AD. However, it is important to note specific genetic predispositions that may result in an increased likelihood and severity of neurodegeneration and dementia.

Conclusions

By reflecting on how the production and exposure to pollutants can harm neurological health, a push for significant legislative changes can be advocated. Air pollutants can adversely impact nearly anyone who inhales them, and chronic production and exposure may lead to devastating diseases. This chapter has explored correlations of ambient air pollutants with Alzheimer's Disease, Parkinson's Disease, multiple sclerosis, and even generalized dementia. Epidemiological studies have shown increases risks for those with chronic exposure,

while controlled mouse models have shown the increased inflammatory and potentially neurotoxic effects of pollutants. Furthermore, air
pollution may affect humans even before birth, with decreased birth
weights and head circumference, and at critical developmental stages,
causing childhood cognitive and behavioral delays. The significance
of air pollution on neurological health cannot be understated, with its
ability to affect a large variety of populations and demographics.
Therefore, now that we know of negative neurological effects by air
pollutant exposure, we should be emphasizing a push for pollutant
control and clean air.

References Cited

Cacciottolo, M., *et al.* 2017 Particulate air pollutants, APOE alleles
and their contributions to cognitive impairment in older
women and to amyloidogenesis in experimental mod-
els.Translational Psychiatry 7.1 (2017): e1022.

Cheng, Hank, *et al.* 2016. Nanoscale particulate matter from urban
traffic rapidly induces oxidative stress and inflammation in olfactory epithelium with concomitant effects on
brain." Environmental health perspectives 124.10 (2016):
1537.

Kioumourtzoglou, Marianthi-Anna, *et al.* "Long-term PM2. 5 exposure and neurological hospital admissions in the northeastern
United States." *Environmental health perspectives* 124.1
(2016): 23.

Malmqvist, Ebba, *et al.* "Fetal growth and air pollution-A study on
ultrasound and birth measures." *Environmental Research* 152
(2017): 73-80.

Mousavi, Sayed Esmaeil, *et al.* "Multiple sclerosis and air pollution
exposure: mechanisms toward brain autoimmunity." *Medical
Hypotheses* (2017).

Ritz, Beate, *et al.* "Traffic-related air pollution and Parkinson's disease in Denmark: a case control study." *Environmental health
perspectives* 124.3 (2016): 351.

Saenen, Nelly D., *et al.* "Recent versus chronic exposure to particulate matter air pollution in association with neurobehavioral performance in a panel study of primary schoolchildren." *Environment International* 95 (2016): 112-119.

Wright, Joshua C., and Yuchuan Ding. "Pathophysiological effects of particulate matter air pollution on the central nervous system." *Environmental Disease* 1.3 (2016): 85

Effects of Climate Change on Vector-Borne Disease Behavior and Transmission

Hogan Marhoefer

Over the last fifty years much of the globe has experienced significant changes in weather because of global climate change and warming. Such changes include increased (and decreased) humidity and precipitation levels, average temperature, and extreme weather events. Much of the science behind global changes in climate is well understood, however despite much research, the impact of climate change on vector populations and disease transmission is not as well understood. Dipteran vectors, such as mosquitos, are known to carry infective pathogens that can be transferred to other species and cause disease. Malaria is the most commonly researched mosquito-borne disease because this protozoan is known to have caused epidemics in countless regions. The aim of many contemporary studies within this scope is to better understand how mosquito populations are impacted by climate variability and extreme weather events, in regions where diseases like malaria are epidemic. However, global climate change has been found to have diverse impacts on local climate depending on the region. In some regions, climate change has contributed to increased precipitation and humidity, which provide a more favorable environment for mosquito populations. Other regions have witnessed reduced precipitation and humidity, which has contributed to lower rates of vector-borne disease transmission. Climate models are commonly employed in these studies to predict potential future climate scenarios. These climate models serve to help estimate potential future precipitation, humidity, and temperature levels. It is impossible to perfectly estimate these values but climate models provide the most accurate framework by which to analyze vector distribution and potential future transmission rates.

The goal of many of these studies is to use climate modeling to best predict the climate in 2050, for example, to expose the regions of the world most vulnerable to future increased transmission rates and epidemics. This scope is especially contemporarily important because of the recent *Zika* virus outbreak. The information derived from the use of climate models (although they are not entirely accurate) to predict future rises in transmission rates and dispersal of vector populations is important for public health officials to consider and is our best chance to reduce transmission rates of vector-borne diseases.

Increased Rate of Malaria Transmission Found Following Severe Flood Event in Western Uganda

It is well-known that malaria is transmitted by mosquitoes and increased transmission rates relate to the success of local mosquito populations. It has also now known that global warming has contributed to increased precipitation and warmer weather in many regions, which provide a more favorable environment for mosquitoe reproduction success (due to increased precipitation). It is apparent that precipitation greatly influences disease transmission in Malaria, however temperature plays an equally influential role in vector-borne disease transmission. As temperatures rise, mosquitoe populations are able to occupy regions of higher altitude as they are no longer restricted by regions of cooler temperatures These factors are of great importance to this study because a significant number of the rural communities in the Kasese District are at relatively high elevation, and have experienced heavy rainfall.

The authors examined the changes in rates of malaria transmission following a significant flooding event in the Western Ugandan Highlands to better unfderstand how transmission rates of malaria are affected by severe changes in local climate patterns. This study is not only pertinent to a better understanding of how severe flooding events can influence transmission rates but also sheds light on how even moderate increases in precipitation levels could potentially impact malaria transmission. Researchers analyzed data from the 12-month period directly before and after the severe flooding event, which they collected from the Bugoye Level III Health Center (BHC). The researchers chose to focus on the *P. falciparum* test positivity rate (or PfPR) and the "incidence rate of severe malaria" to asses the impact of the 2013 flooding event on malaria transmission.

The study found that the test positivity rate had increased from 25.8% to 39.4% from the preflood to the postflood period. The analysis of PfPR by month illustrated that the most significant increase had taken place during the month of November, noteably a low transmission period. In addition, researchers found that postflood PfPRs were significantly higher in villages that bordered flood-affected rivers. It was found that flood-effected rivers were present in all of the eight villages where significant changes in malaria risk had been seen, suggesting that villages in closer proximity to flooding rivers are more vulnerable to malaria transmission. This article highlights the need for more research to examine the disease control implications of other severe weather events as well as developing a better understanding of the magnitude of effect on transmission that could be seen by mere changes in precipitation levels resulting from global climate change.

Assessing the Impact of Climate Change on Human Infectious Diseases

With today's great wealth of knowledge in science concerning the our climate, it has been widely accepted that global warming and global climate change has been amplified by anthropogenic activities. The scientific community has prodecued a large quantity of literature that contributes to the aforementioned wealth of knowledge concerning the science of climate change, however many of the subtle impacts of global climate change are misunderstood. The impact of climate change on disease control and the transmission of vector-borne diseases are, unfortunately, areas of study that do not see as much attention. Fortunately, however, much has been done in recent years in attempt to fill in some of the unknown gaps in this scope of science.

Wu *et al.* (2016) conducted a review of the impact of climate change on infectious disesases. The researcers address some of the principle concerns of a changing environment in terms of disease control. Increased precipitation, humidity and temperature are often associated with global climate, and these factors can also impact the transmission of a number of vector-borne diseases. The authors explain that temperature changes resulting from climate change can influence pathogen life-cycles, and in certain cases this can reduce the incubation period of these pathogens. Rising precipitation levels have similarly troubling effects on disease transmission. Increased precipitation levels can result in higher rates of water-borne disease transmission by increasing the number of fecal pathogens and providing more areas of stagnant water for the accumulation of fecal pathogens. The authors also make important note of the routes by which climate change is known to influence vector-borne disease transmission. Not only does increased rainfall contribute to a wider range of potential breeding habitat for many pathogen-bearing dipteran species, but an increase in temperature can also broaden the spatial-temporal distributions of many pathogen-carrying insects, which promotes their existence in new envronements (usually at higher altitudes).

It is also important to note the relationship between extreme weather events and infectious disease control. Global climate change has increased the severity of extreme weather events, as well as the total number of individual extreme weather events, and many studies have confirmed that there is a relationship between extreme weather events and increased infectious disease incidence. Water- and vector-borne diseases appear to be influenced the most by severe weather events, and although infrequent, food-borne disease transmission is occasionally impacted by severe weather events. The information presented in this review paper provides a very comprehensive illustration of how climate change can influence the behaviors of a number of different infectious diseases, and provides evidence that increased vulnerability to infectious diseases commonly follows extreme weather events. It is clear from this paper that a large amount of scientific literature supports the idea that global climate change influences the spread of disease. It becomes additionaly apparent that certains regions of the globe are unprepared and will be unable to respond to the continuous effects of global warming. Thus, this study is particularly relevant to international healthcare and disease control practices, as the need for a competent solution becomes evident. The authors offer a thorough and detailed understanding of how climate change influences infectious disease behavior, and, hopefully, this article contributes to a more hollistic interpretation of the many ways climate change can influence this fragile planet.

An Analysis of the Impacts of Global Warming on Infectious, Waterborne Diseases

It is well known that global warming has contributed to large-scale changes and many professionals are concerned about the indirect impact of global warming on disease transmission. Some of the climatic changes from global warming include increased precipitation, humidity, and temperature, all of which have been shown to influence the behaviors of certain infectious diseases. The authors explain that climate change and severe weather events can promote waterborne disease transmission. By interfering with stable hydrologic cycles. The authors also emphasize global warming's impact on the number of severe weather events and reveal that severe weather events have grown in scale and number with global warming.

Because of climate change the incidences of waterborne infectious diseases have increased, and O'Dwyer *et al.* explain that climate change is the reason behind these changes in disease behaviors. Waterborne disease incidence is the most concerning vector of infectious disease transmission that the authors focus on in this study. In regions of the world where there is a lack of water-management infrastructure this problem could have especially dramatic effects. The authors explain that the incidence of Cholera, Giardiasis, Typhoid, and many other infectious diseases increase from climatic changes. Increases in temperature have been revealed to directly contribute to transmission as temperature impacts certain diseases growth rates. Many incidents of bacterial diseases have been shown to increase in areas where management of waterborne pathogens is poorly controlled.

This article is important in many aspects health, disease control and disease management. The world is changing and it is imperative that we gain a better understanding of how climatic changes impact other areas of our life. It is accepted that global warming has changed our environment, and this article properly articulates many of the potential concerns regarding waterborne disease control.

El Niño Found as a Driving Force Behind Widespread Transmission of the Zika Virus

2016 was a year of many ups and downs. In 2016, we witnessed the effects of an unexpected virus when the Zika virus (ZIKV) emerged in South America. Caminade *et al. (2016)* developed a "global R0 model" for ZIKV that includes the relevant vector species (*Ae. albopictus and Ae. aegypti)* and adjusts for dynamic climate change. The model pulls data from recently published studies concerning estimates of global abundance and "global observation-based historical climate data" to derive transmission potential (or R0) estimates that illustrate transmission risk. The authors show that tropical regions have the highest mean R0, per the model. However, the model also suggests that the average R0 of ZIKV is greater in South America any other region in the world. It is also apparent that southeastern Asia is particularly susceptible to ZIKV. Alike other articles in this topic, Caminade *et al.* point out that environmental conditions are suitable for ZIKV transmission to occur over a wider geographical range". The structuring of the model allows for the analysis of respective ZIKV contributions and the authors found that *Ae. aegypti* is responsible for >90% of ZIKV transmission risk in the tropics. However-er, the model predicts that *Ae. albopictus* is the main vector responsi-ble for transmission risk in temperate, northern regions of the globe.

The authors also learned from the model that extreme weath-er events tend to "favor epidemics" which could potentially exacer-bate transmission rates. Caminade *et al.,* state that the 1998 – 99 and 2015 *El Niño* events witnessed a great increase in ZIKV, largely due to the increase in regional precipitation, humidity, and temperature levels. In the discussion, the authors acknowledge that the combined effects of warmer temperatures from global warming and warmer re-gional temperatures from *El Niño* influenced vector-borne ZIKV dis-ease transmission risk in South America. The authors reiterate that *Ae. aegypti* impose a greater risk to worldwide ZIKV transmission than *Ae. albopictus.* This information has the potential to influence international disease management strategies, and can inform infra-structural decisions. However, I do not fully understand how the model accounted for regional public health and infrastructure measures in place, as certain regions have better public health systems in place to combat the spread of Zika virus.

Under Climate Change, Ticks and Dipteran Vectors Display Different Disease Behaviors

Much of recent literature pertaining to vector-borne disease transmission has focused on mosquito-borne diseases, however Og-den and Lindsay (2016) maintain that this narrow scope leaves cer-tain questions unanswered. In their review, Ogden and Lindsay (2016) assess how climate change may impact dipteran vectors, such as mosquitoes, compared to hard-bodied ixodid-borne (tick) diseases. Vector-borne disease transmission is impacted by temperature varia-bility under the current, changing climate, as higher temperature is conducive to greater population mobility in these vectors. This in-formation is supported by data in the form of graphic representations of factors such as annual mean temperature and precipitation change.

In their comparison of multivoltine dipteran and ixodid tick disease vectors, Ogden and Lindsay (2016) explain that contrasting life histories, life cycles, and population behaviors impact how such vectors can respond to changes in the climate. Table 1 summarizes the attributes potentially impacted by climate change and explains how it could impact the spread of the disease. Particularly interesting is the fact that ticks have a much longer life cycle than mosquitoes, and as they do not require water for breeding they are less reliant on specific habitat for reproduction. In Canada, Lyme disease and WNV (West Nile Virus) emerged together, which allows us to more easily compare their vector habits.

The authors reveal the number of reported cases of Lyme disease and West Nile Virus to compare the trends in disease distribution. Although certain years (particularly 2003 and 2007) have witnessed increases in WNV, the number of cases of Lyme disease is apparently steadily growing, which is concerning. It is explained that the key difference between dipteran and tick vectors is the speed at which their populations can respond to short-term changes in weather. It is concluded that dipteran-borne diseases have the capacity to 'boom and bust' over short time periods, whereas ticks cannot respond in this way.

Ethiopia Predicted to be Vulnerable to El Niño-Related Malaria Epidemic

It has been widely accepted that global warming has an immediate impact on various aspects of public health. Bourma *et al.* (2016) provide an interesting article that concerns public health and the direct impact of El Niño on malaria cases in Ethiopia. Global climate change has increased the severity of the current El Niño, principally in terms of precipitation and temperature levels. Such environmental changes provide a favorable environment for many dipteran species, and usually lead to increased levels of dipteran-borne diseases, like malaria.

In Ethiopia between August 2016 and July 2017, Bourma *et al.* (2016) predict an increase in malaria risk of around 70%. The authors explain this to be a result of increased local mean surface temperatures during the months of December 2015 and March 2016. It is also possible that other areas could see an increase in malaria risk because of the potential altitudinal shifts in the habitats of certain dipteran species as temperatures increase in those regions. It is important to note that many areas *can* cope with the increase in malaria risk with certain 'preventative operations' including rapid diagnostic tests and providing antimalarial drug supplies to locals. The problem arises in areas of higher altitude where increasing higher average temperatures provide a suitable habitat for dipteran species and the emergence of disease.

A Lot to be Learned from the Successful Elimination of Malaria in Ecuador and Peru

Malaria is a prominent epidemic vector-borne disease in many regions of the world today. Many of these areas have had much difficulty in lowering local transmission rates, especially considering the indirect influence on vector-borne transmission rates by climate change. However, some areas have been successful in eradicating malaria from their population. A case study was performed in the *El Oro Province,* in Ecuador, and the *Tumbes Region* of Peru, where a surge in malaria transmission began in the mid 1980s and lasted until the early 2000s. Krisher *et al.* (2016) examine the factors that contributed to the successful elimination of malaria in the region, and hope to inform and educate future public health professionals there.

The authors analyze what interventions played key roles in eliminating malaria from this region. Krisher et al (2016) reveal the importance of health protocol overlap amongst public and private clinics within the region, as well as "comprehensive and standardized treatment, monitoring, and follow-up of lab confirmed cases." It was also important for public health professionals in this area to consult other research to combat the influence of climate change on malaria transmission. These steps allowed health professionals in the region to better anticipate changes in transmission rates due to climate change and weather fluctuations. Among other preventative practices, health officials employed strategies that targeted the most vulnerable population, including pregnant women and children under 5.

Disease control in border regions is very difficult to manage. As both people and goods are in constant movement, vectors can find new hosts or establish themselves in new regions. Therefore, other regional partnerships to support malaria control in the border region were one of the most important factors in eliminating malaria from the region. This collaboration proved to effectively suppress malaria in these regions of Ecuador and Peru, leading to complete elimination of the disease in El Oro and Tumbes in 2011 and 2012, respectively.

Temporal Influences of El Niño and Weather on Dengue in Sri Lanka

The principle health concern in Sri Lanka is dengue, a mosquito-borne virus. Climate change is known to have an impact on the population behaviors of mosquito species can contribute to changes in dengue transmission. Warmer temperatures and increased precipitation, because of climate change, allow dipteran vectors a greater range of potential habitat and provide more standing water for potential breeding sites.

The authors explain that the overall relative risk of dengue increased as temperature rose (Liyanage, 2016).

Sri Lanka's climate is dominated by monsoon seasons, which provide the perfect habitat for mosquito population growth and breading. The authors reported a total of 7412 cases of dengue from 2009 to 2013. According to the article, rainfall (when increased from 100 mm to 250 mm) significantly increases the relative risk of dengue. In addition, increasing temperature was found to have a direct, linear effect on relative dengue risk. In Sri Lanka, where trends of increased precipitation and elevated temperature are seen, the risk of increased dengue transmission is very concerning. The most significant "positive association was seen at weeks 8 – 10 for cumulative rainfall above 300 mm per week." (Liyanage, 2016). But what does this mean to the public health management? The authors explain that rainfall has greatest ability to influence rates of dengue transmission, especially in an area where monsoons are common.

Conclusions

In conclusion, the research concerning the changes in disease behavior that result from anthropogenic changes in the climate display the resilience of certain diseases in a changed climate, and the public health implications of climate change affecting disease vectors. Countless authors have contributed research considering vector-borne diseases and the effect climate change has on mosquito-borne diseases. From this research, it has been found that many dipteran species are vulnerable to variable weather. Dipteran species, like mosquitoes, prefer humid habitats with bodies of stagnant water to breed, as we know. However, researchers have found that changes in the climate have the potential to assist mosquito populations and, thus, increase mosquito-borne diseases. Extreme weather events, such as a flooding event (as seen in the study conducted in the Western Ugandan Highlands) have a direct effect on mosquito populations and malaria transmission rates. This research found that the rural communities that lay closest to rivers experienced the greatest increases in the total number of cases of malaria. Researchers affirm that specific regions have experienced increased precipitation because of global climate change, however, some regions have witnessed the opposite. Some areas have become even drier, experiencing much less rainfall, and as a result have witnessed a reduced number of malaria cases. In those regions, increased average temperature has been found to extend existent mosquito populations geographic range. In cases like this, mosquito populations have extended into areas of higher altitude where lower temperatures have previously restricted the population's movement into these regions. Researchers have found that communities at higher altitudes are experiencing increased rates of malaria transmission because mosquito populations are now able to move to new, high-altitude regions. It is clear from the research that climate change directly impacts dipteran species' populations, and thus indirectly impacts mosquito-borne disease transmission rates. Other vector species, like ticks, which carry diseases such as Lyme disease, have been found to be less vulnerable to climate change and variability. Researchers have attempted to predict climate change using climate models, which have a degree of error as these models only provide estimations. It is important to understand climate models as thoughtful predictions and consider that they are not always accurate. This is important because the authors address the need for continued practice of climate modeling so that accuracy can be improved, and more precisely estimate future climate scenarios. This information is important from a public health standpoint, however, because the ability to accurately predict future climate scenarios could potentially allow public health officials a new understanding of how to predict and contain weather-related changes in transmission rates.

References Cited

Bouma, M. J., A. S. Siraj, X. Rodo, and M. Pascual, 2016. "El Niño-based malaria epidemic warning for Oromia, Ethiopia, from August 2016 to July 2017." Tropical Medicine & International Health 21.11: 1481–488.

Caminade, Cyril, and Joanne Turner, 2016. "Global risk model for vector-borne transmission of Zika virus reveals the role of El Niño 2015." Proceedings of the National Academy of Sciences 114.1: 119 – 24.

Krisher, Lyndsay K., Jesse Krisher, Mariano Ambuludi, Ana Arichabala, Efrain Beltrán-Ayala, Patricia Navarrete, Tania Ordoñez, Mark E. Polhemus, Fernando Quintana, Rosemary Rochford, Mercy Silva, Juan Bazo, and Anna M. Stewart-Ibarra, 2016. "Successful malaria elimination in the Ecuador–Peru border region: epidemiology and lessons learned." Malaria Journal 15.1: 20–34.

Liyanage, Prasad, Hasitha Tissera, Maquins Sewe, Mikkel Quam, Ananda Amarasinghe, Paba Palihawadana, Annelies Wilder-Smith, Valérie Louis, Yesim Tozan, and Joacim Rocklöv, 2016. "A Spatial Hierarchical Analysis of the Temporal Influences of the El Niño-Southern Oscillation and Weather on Dengue in Kalutara District, Sri Lanka." International Journal of Environmental Research and Public Health 13.11: 1087.

O'Dwyer, Jean, Aideen Dowling, and Catherine Adley, 2016. "The Impact of Climate Change on the Incidence of Infectious Waterborne Disease." Urban Water Reuse Handbook 1832.1: 1017–026.

Ogden, Nick H., and L. Robbin Lindsay, 2016. "Effects of Climate and Climate Change on Vectors and Vector-Borne Diseases: Ticks Are Different." Trends in Parasitology 32.8: 646–56.

Ross Boyce, Raquel Reyes, Michael Matte, Moses Ntaro, 2016. Severe Flooding and Malaria Transmission in the Western Ugandan Highlands: Implications for Disease Control in an Era of Global Climate Change. The Journal of Infectious Diseases 214 (9): 1403–1410.

Wu, Xiaoxu, Yongmei Lu, Sen Zhou, and Lifan Chen, 2016. "Impact of climate change on human infectious diseases: Empirical evidence and human adaptation." Environment International 86: 14–23.

About the Authors

The authors of this book are students at the Claremont Colleges. The book is a work product of Biology 159: Natural Resources Management taught by Emil Morhardt in the W.M. Keck Science Department of Claremont McKenna, Pitzer, and Scripps Colleges. Each student picked a topic, did a literature search, and selected eight papers written within the past year that exemplified the state of the science.

Their task was to write journalistic summaries capturing the essence of the papers but eschewing technical terms to the extent possible—to become, in effect, science writers. The summaries were due weekly and were returned with editorial comments shortly thereafter. The chapters are compilations of the individual summaries with additional introductory material and a brief conclusion.

The editor is Roberts Professor of Environmental Biology at Claremont McKenna, Pitzer, and Scripps colleges. He remembers how difficult it is to learn to write and appreciates the professionalism shown by these students.

Index

Index